Isaac Winter Heysinger

The source and mode of solar energy throughout the universe

Isaac Winter Heysinger

The source and mode of solar energy throughout the universe

ISBN/EAN: 9783744741644

Printed in Europe, USA, Canada, Australia, Japan

Cover: Foto ©berggeist007 / pixelio.de

More available books at **www.hansebooks.com**

THE
SOURCE AND MODE
OF
SOLAR ENERGY

THROUGHOUT THE UNIVERSE.

BY

I. W. HEYSINGER, M.A., M.D.

ILLUSTRATED.

PHILADELPHIA:
J. B. LIPPINCOTT COMPANY.
1895.

COPYRIGHT, 1894,
BY
I. W. HEYSINGER.

All rights reserved.

CONTENTS.

	PAGE
INTRODUCTION	7

CHAPTER I.
STATEMENT OF THE PROBLEM OF SOLAR ENERGY ... 17

CHAPTER II.
THE CONSTITUTION AND PHENOMENA OF THE SUN ... 39

CHAPTER III.
THE MODE OF SOLAR ENERGY ... 70

CHAPTER IV.
THE SOURCE OF SOLAR ENERGY ... 96

CHAPTER V.
THE DISTRIBUTION AND CONSERVATION OF SOLAR ENERGY 139

CHAPTER VI.
THE PHENOMENA OF THE STARS ... 162

CHAPTER VII.
TEMPORARY STARS, METEORS, AND COMETS ... 187

CHAPTER VIII.
THE PHENOMENA OF COMETS ... 210

CHAPTER IX.
INTERPRETATION OF COMETIC PHENOMENA ... 225

CHAPTER X.

THE RESOLVABLE NEBULÆ, STAR-CLUSTERS AND GALAXIES 237

CHAPTER XI.

THE GASEOUS NEBULÆ 253

CHAPTER XII.

THE NEBULAR HYPOTHESIS: ITS BASIS AND ITS DIFFICULTIES 268

CHAPTER XIII.

THE GENESIS OF SOLAR SYSTEMS AND GALAXIES 282

CHAPTER XIV.

THE MOSAIC COSMOGONY 308

CHAPTER XV.

CONCLUSION. THE HARMONY OF NATURE'S LAWS AND OPERATIONS 341

REFERENCE INDEX OF AUTHORITIES CITED 349
CLASSIFIED INDEX OF SUBJECT-MATTER 353

LIST OF ILLUSTRATIONS.

	PAGE
Figs. 1 to 8. Types from nature, illustrating development of a solar system from the attenuated matter of space .... *Frontispiece*.	
Fig. 9. A typical sun-spot...............	57
Fig. 10. Structure of the sun, analytical illustration of	60
Fig. 11. Electrical polarities of sun and planets	82
Fig. 12. Ideal view of the generation and transmission of planetary electricity.................	89
Fig. 13. The aurora borealis, view of	91
Fig. 14. Diffused brush discharge of an electrical machine	91
Fig. 15. Planetary generation and transmission of electrical energy to the sun, analytical illustration of	101
Fig. 16. Gradual discharge of electricity from one conductor to another in a partial vacuum	103
Fig. 17. Sudden electrical discharge through the atmosphere ...	103
Fig. 18. Position of planets with reference to the generation of sun-spots; maximum and minimum of electrical action	108
Fig. 19. Analysis of a typical sun-spot	112
Fig. 20. Retardation of sun-spots in their travel across the solar face; development to the rear and recession in front	114
Figs. 21 and 22. Complex lines of planetary electrical action upon the sun produced by the inclination of the solar axis to the plane of the ecliptic..................	120
Figs. 23 to 29. Examples of electrical repulsion: Fig. 1, similarly electrified pith-balls; Fig. 2, the electrical windmill; Fig. 3, repulsion of a flame; Fig. 4, self-repulsion around a conductor; Fig. 5, attraction between opposite and repulsion between similar electricities; Fig. 6, mutual repulsion between similar + electrospheres of the earth and the moon; Fig. 7, mutual repulsion between the similar — electrospheres of sun and comet	124
Figs. 30 to 34. Spectra of solar light, incandescent sodium and calcium, and the absorption and bright-line spectra of hydrogen gas	155
Figs. 35 to 37. Reversal and neutralization of spectroscopic lines of hydrogen in the light of a variable star like Betelgeuse	160

LIST OF ILLUSTRATIONS.

	PAGE
Fig. 38. A double-sun nebula in process of development into a solar system	164
Fig. 39. Double stars with complementary colors, interpretation of the phenomena of	167
Fig. 40. A solar system which would explain the regular variability of the star Mira	178
Fig. 41. Lineal nebula in Sobieski's Crown which has been affected by currents in the ocean of space	189
Figs. 42 to 45. Four stages in the phenomena of a new or temporary star, a "star in flames;" reversal of the hydrogen lines in its spectrum	196
Figs. 46 and 47. Illustration of repulsion of the tail of a comet by the similarly electrified solar electrosphere; comparison with similar repulsion in a vacuum-chamber experiment	211
Figs. 48 and 49. The electroscope, and mutual electrical repulsion in a bundle of dry straws	225
Fig. 50. Experiment with a candle and currents of air from between two disks, illustrating the radial semi-rotation of a comet's tail during perihelion	230
Figs. 51 to 54. Four non-systemic gaseous nebulæ: Fig. 1, crab nebula; Fig. 2, dumb-bell nebula; Fig. 3, lineal nebula in Sobieski's Crown; Fig. 4, Catherine-wheel nebula. The latter illustrates the formation of a planetary nebula with a hollow center, or else dispersion into the elements of space again	263
Fig. 55. Great spiral nebula in Canes Venatici and a small adjacent nebula affected thereby	273
Figs. 56 to 59. Four gaseous nebulæ in process of development into solar systems: Fig. 1, divergent spiral; Fig. 2, later stage of a similar spiral; Fig. 3, subsequent stage of rupture of the nearly circular convolutions of a similar nebula; Fig. 4, the same stage in the development of a solar system with a double sun	279
Fig. 60. Nucleated planetary nebula, showing its external ring split and held apart, in part of its circumference, by electrical repulsion	288
Fig. 61. Divergent spiral nebula on cover of book.	

INTRODUCTION.

THIS work is not presented to the reader as a treatise on astronomy, although the different phenomena pertaining to that splendid science are reviewed with some detail, and the established facts bearing upon the subjects discussed are briefly cited in the very words of the great writers upon whose authority they rest. A considerable experience in chemistry, electricity, and the other allied physical sciences long since convinced the author of this work that some simple and uniform principle must control the production of the physical phenomena of astronomy,—some general law capable of being extended in its application to the widest, as well as applied to the narrowest, limits of that science. Knowing the absolute certainty of a magnetic and electrical connection between the sun and the earth, as evidenced by the reflected energy of sun-spots, auroras, etc., and that no known cause except electricity could account for some, at least, of the cometic phenomena, it seemed that any comprehensive law must at all events include this mode of energy as an effective cause, and that if the law be uniform in its application, it must equally exclude all others which may be either antagonistic or not necessary. A careful investigation was therefore made of those less generally known principles concerned in the generation and transformations of electrical energy, in order to determine the sufficiency or insufficiency of this agency in the grander operations of nature (for, of course, mere currents of electricity could play no part in these phenomena), with the result that every line of research led irresistibly to the conclusions presented in this work. These investigations, specifically directed, at first, to the source and mode of the solar energy of our own system alone, were found to be equally applicable to others, and were successively extended to the whole sidereal, nebular, and cometic field, and finally to space itself, for all the phenomena of which it seemed to furnish an adequate and harmonious interpretation. The fact, when once demonstrated, that the true source of *solar* energy is not to be found in the sun itself, but in the potential energy of space, served as a guiding principle, and, by its continuously extended application, was found to cover perfectly the source and mode of all solar energy. Every step of the investigation has been based on the established facts of science and the observations of eminent astronomers as laid down by the best authorities; and the quotations herein made from their works are full and fair, and are properly credited in every case, and taken from books easily accessible to the general reader. It is hoped that further attention may be directed to this field of research by far more capable investigators than the author of this work, so that systematic astronomy may no longer bear the reproach that it is largely an empirical science, but that it may henceforth be based upon rational and comprehensive principles, capable of universal extension and of general scientific application.

The authorities cited in this work include many illustrious names: Proctor, Tyndall, Helmholtz, Langley, Huggins, Newcomb, Young, Flammarion, Balfour Stewart, R. Kalley Miller, Herschel, Nichol, Lord Rosse, Urbanitsky, Crookes, Fraunhofer, Ball, and many others, all of whom are known throughout the world as among the master minds of science. From them we have drawn the rich stores of knowledge of the phenomena with which this work deals, and which we have so fully and freely cited, as the basis of the splendid superstructure which astronomy to-day reveals. No one will venture to controvert the statements of fact made by these eminent men, and, where conflict of opinion has arisen among them, we have quoted all parties, so that the reader can form his own conclusion, in each case, for himself. So diverse, apparently, are the phenomena reviewed that they present the aspect of a great picture-gallery, in which the paintings totally differ from each other in subject, in treatment, and in origin, their only common qualities being those of grandeur and fidelity to truth and to the principles of art. But they are not merely paintings, they are the moving panorama of creation, and, diverse as they may appear, they will be found to show the same " handling," which reveals the same universal artist; they have, in truth, a common mode of development and a common principle of construction, obscure as these may seem to be.

For thousands of years "Natural History," so called, was studied and taught; zoölogy was a well-known science far back in old historic times. But it was left for modern biological research to turn from these fixed and fully-developed forms of life, and go back to trace their primal development through what is now the science of embryology, and thus we have learned that nature traverses the same paths in forming a man as in producing a frog or a bird. The process is carried further along in one case than in another, but the lines of development are almost identical; and the tracing out of these common lines and their subsequent divergencies has shed a flood of new light upon these dark and hitherto unknown places, so that we are now fairly on the true highway of physical life at last. When adult forms were alone compared, animal with animal, no common ground of origin or development could be discerned; nature was believed to work by "special creations," and vast cataclysms were devised to utterly destroy the organic life of one terrestrial epoch after another, leaving a few hardy accidental survivors, or "types," perchance, to trace back their lines of descent beyond such periods of cyclical destruction. All this is now changed, and these views, so recently held and taught, have been abandoned forever, and continuously operative natural processes of development, modified by environment and heredity, have taken their place, and biology now has a future as well as a past. And so it must be with the less complex, but far more extended, creations and transformations in the vast fields of astronomical science with which this book is concerned. Hitherto we have here, too, dealt with "special creations" and cataclysms; henceforth we must follow the uniform and eternal laws of progressive development.

Among the multitude of hitherto unsolved problems of astronomy we may enumerate the following: Why sun-spots travel faster around the sun when near his equator than when more distant from it. The physical causes of sun-spots, faculæ, and solar prominences. Why the number and size of sun-spots seem to affect terrestrial magnetism. The rational interpretation of the eleven-year and the long sun-spot cycles. The origin of the aurora borealis. The causes of the periodicity of regularly variable stars. How to explain, in accordance with the nebular hypothesis, why Algol and its companion, which are not greatly different in mass and volume, and both obviously gaseous, should so differ in character, one being a bright sun and the other a dark planet. Whether there are great, compact, but dark bodies, comparable to suns and planets in magnitude, and unconnected with any solar system, floating about in space. Why double and multiple stars are so frequently of contrasted or complementary colors. Why regularly variable stars are longer in decline than in growth of brilliancy, since such decline is no criterion of loss of heat, but rather the reverse. Why the sun and fixed stars have atmospheres largely composed of free hydrogen, and the planets have atmospheres of free oxygen and nitrogen. Why a small and sometimes even scarcely visible star occasionally is seen to suddenly blaze up, in a few hours, to hundreds of times its normal brilliancy, and then far more gradually fade, through months and years, back to its former state, in which thenceforth it continues to maintain its original lustre. Why comets, when they have tails, always project these appendages radially from the direction of the sun. How to account for the presence of cyanogen, and how for the absence of oxygen and the constant presence of hydrocarbon vapors around the nuclei of comets. Why some comets split up into separate comets and others sometimes show multiple tails. Why comets, when they pass around and behind the sun, in some cases reappear shorn of their splendor and in other cases with their splendor greatly enhanced. Whence comets are derived, where is their permanent abiding-place, and how did they originally reach those distant regions which they occupy before entering our system, if merely the *débris* left behind from contraction of the mass of plasma out of which our solar system is supposed to have been formed. Why so many of the irresolvable nebulæ present the appearance of divergent spirals of many different forms. How to account for the annular nebulæ with hollow centers and for those partially-completed planetary nebulæ, so called, which afterwards appear to retrograde into diffused gaseous nebulæ again or gradually disappear. What is the ultimate constitution of interstellar space? Have the fixed stars planetary systems like our own, or not? Must they have such, or merely may they have? What principle of conservation of energy is it possible to apply to the vast quantities of light and heat which constantly disappear in the interstellar realms of space? How to account for this enormous emission of solar energy during the long period of time requisite for the development of the earth during its past geological ages. How to explain why the moon always presents the same face to the earth. Why, if the law of gravity prevails there, there are no visible traces of atmosphere or moisture in the moon. What is the basic principle on which depends the ratio of mean planetary distances, 0, 3, 6, 12, 24, etc., *always plus 4?* What is the origin of the planetary satellites and the cause of their irregular distribution, and what the origin of Saturn's rings? How was the belt of asteroids formed between Mars and Jupiter? Why is the orbit of Neptune relatively compressed against that of Uranus? Why is the mass of Neptune out of its proper proportion compared with those of Jupiter, Saturn, Uranus, and Neptune in a diminishing series? What is the rational interpretation and what the origin of the sun's corona and the cause of the coronal streamers?

There are many other problems equally difficult which are encountered in the study of this noble science, but the above are surely sufficiently striking. Any complete interpretation of these various phenomena, even singly, would seem to be an important step in advance; then how much more so if the explanation of one and all of these is to be found in a single, all-embracing cause, a few simple and uniformly operative principles, as unquestionably operative here as in the other fields of science to which they pertain, and which, once thoroughly comprehended and rigidly applied, will be found to elucidate all the multifarious phenomena of sidereal space so clearly and precisely that any intelligent observer and reasoner can determine each question finally for himself, and solve not only these, but all the other astronomical problems and paradoxes which have from time to time arisen? It is not to be understood that this sublime science and these illimitable realms are to be laid off with the metes and bounds of a farmer's meadow, for all the lines of the different sciences are linked together at a thousand points, but that the operative principles which nature constantly employs once firmly grasped, the intricacy of each series of phenomena encountered will become gradually lessened, link by link, as observations and deductions are more closely and rationally made along these well-established lines of research, instead of here and there, empirically, and at hap-hazard, as has been the only method hitherto possible to pursue. When the relatively few fixed principles which control the operations of nature in the field of astronomy are thoroughly comprehended, for on this vast panorama she lays her colors with a heavy brush, we can study her phenomena and interpret her processes even more readily than the kindred sciences have enabled us to do in the adjacent fields of biology, wherein the splendid achievements of less than a quarter of a century past have not only aroused the interest and enthusiasm of the world, but already point the way to still grander triumphs yet to come.

THE SOURCE AND MODE
OF
SOLAR ENERGY.

CHAPTER I.

STATEMENT OF THE PROBLEM OF SOLAR ENERGY.

In endeavoring to present a new and rational interpretation of the source and mode of solar energy, based upon the established principles of recent science, it becomes necessary to briefly cite the facts bearing upon the problem to be solved and the authorities for their support, as well as to describe concisely the different hypotheses at present in vogue, and to point out the well-established insufficiency of these theories, one and all, to account for or explain the difficulties encountered, and which so far have remained as an unsolved enigma. And this problem of solar energy is the grandest and most important question of all physics, for upon the light and heat of the sun depend all physical life and its consequences, animal and vegetable, past, present, and future. If within finite time, and relatively, compared with the enormous vistas of the past, a very brief time, this source of energy is to cease, and our whole system be involved in darkness and death, such darkness and death must be eternal; for the dead sun in his final stage of condensation will be as fixed and unchangeable as the operation of eternal laws can make it, and henceforth there can be no revival or reversals, no turning back of the hand upon the dial, while the laws of nature continue; and outside the uniform operation of the laws of nature there is no source, or mode, or continuance of solar energy conceivable. It is true that when our system shall have run down to its culmination in death, other present systems may continue for a time to exist and new ones spring into being; but these, too, must inevitably follow the same course, and likewise end in eternal darkness, until finally the great experiment of creation shall have ended in eternal failure. The changes we see in progress around us, however, are not of this nature. The individual dies, but the forces which gave life and strength to the race persist, and others will take his place, and the same forces will continue to operate with constant renewals, since we draw our light and heat and life from without; but in the death of suns and their attendant planets there is no analogous process, for such suns are constantly expending their enormous energies in the support of life external to themselves, and only the smallest part of this energy, even, can ever be utilized by themselves or by other suns or planets under any mode of interpretation now in vogue, the boundless realms of so-called inert and empty space receiving the same proportionate quota of light and heat as the almost microscopic points in the sky which constitute the suns and systems we see, and practically all, or nearly all, of this enormous energy is an absolute dead waste; so that whether receiving new supplies from a constant rain of adjacent meteor streams, or from the gradual contraction of the solar volume, the vast realms of space are the useless recipients of what can never return to the sun again, and, of course, in such case the inevitable end can be predicted; for contraction of volume, with a given mass, must have an effective limit, and meteoric aggregation must also find an effective limit, if the planets are not to be thrown out of place as they continue to revolve around the sun.

All accepted theories begin with a primordial impulse, the energies of which are of necessity constantly frittered away and wasted, until finally all light and heat and life must cease to exist, and that at a stage in which no further impulse can ever be given, since the whole universe will have passed through every possible stage of degradation down to the final one of universal and eternal death. And yet this is the best that science has to suggest; the only comfort offered us is that it will not happen in our time, and so, "after us the deluge." The nebular hypothesis, so called, of Laplace, has required much modification, in the light of more recent science, but the essential principles of this theory are still generally accepted, for they fairly well account for the primal connection of the sun and planets, and the position of the central sun within, with the orbital and rotational planetary movements, as no other theory has yet done. By this theory the limits of our solar system were once occupied by an attenuated gaseous nebula containing within itself all the matter which now forms our solar system. This great nebular mass, primordially assumed, was given by gravity a slow but gradually increasing rotation upon its center; the force of gravity acted more strongly upon this rotating body as it contracted, so that rings of nebulous matter were successively thrown off, which coalesced into single masses and these finally into planets. These planetary globes themselves, as they coalesced and contracted, left behind or threw off rings of their outer matter, which, in turn, became moons, and finally our solar system with its central sun was evolved as we now see it; development continued, the planets cooled and condensed, life appeared when the conditions became suitable, and the original progressive condensation of the central mass—the sun—still continuing, the evolution of light and heat continues, and will continue in a correlative degree. As our moon has passed, apparently, beyond the stage of life, and is cold, airless, waterless, and dead, so will the earth pass; and the larger planets, such as Jupiter and Saturn, which have not yet reached the life stage of condensation, are still hot, but they, too, will pass through the present stage of the earth, then through that in which the moon now is; and the central sun, still glowing, but more and more dimly, will itself pass through the stages in which Jupiter and Saturn now are, then through that of our present earth, and finally into that of the moon, long before which time the emission of all light and heat will have ceased from the sun to its encircling planets, and finally the sun itself will sink into eternal frigidity, and all its store of light and heat will have been dissipated into boundless space, and the possibility of anything resembling what we know as life will have been forever extinguished. In considering the question of the sun's energy, the author of the article "Sun," in Appleton's Cyclopædia, says, "How to account for the supply of the prodigious amount of heat constantly radiated from the solar surface has offered a boundless field of hypothesis. One conjecture is that the sun is now giving off the heat imparted to it at its creation, and that it is gradually cooling down (1). Another ascribed it to combustion (2), and a third to currents of electricity (3). Newton and Buffon conjectured that comets might be the aliment of the sun (4); and of late years a somewhat similar theory (first broached by Mr. Waterston in 1853) has been in vogue,—viz., that a stream of meteoric matter constantly pouring into the sun from the regions of space supplies its heat, by the conversion into it of the arrested motion (5). As the sun may, indeed, derive a small amount of heat from this cause, it deserves more attention than previous conjectures. But conjecture and hypothesis may be said to have given place to views which claim a higher title, as it is now becoming generally recognized, in accordance with modern physical theories of heat, that in the gravitation of the sun's mass toward its center, and in its consequent condensation, sufficient heat must be evolved to supply the present radiation, enormous as this undoubtedly is. It appears to be susceptible of full demonstration that a contraction of the sun's volume of a given definite amount, which is yet so slight as to be invisible to the most powerful telescope, is competent to furnish a heat-supply equal to all that can have been emitted during historical periods. According to this theory, then (which is largely due to the development by Helmholtz of Mayer's great generalization), the sun's mass remains unaltered, and its temperature nearly constant, while its size is slowly diminishing as it contracts; so slowly, however, that the supply may be reckoned on through periods almost infinite as measured by the known past of our race, and which are in any case to be counted by millions of years (6)." To these must be added the hypothesis of Dr. Siemens, fully described in Professor Proctor's "Mysteries of Time and Space." This ingenious theory, in brief, is that the rotation of the sun on its axis causes a suction in the manner of a fan, at the poles, and a tangential projection, at the equator, of a disk-like stream of gaseous matter into space. The light and heat of the sun, dispersed through space, slowly but continuously act upon the compound gases with which space is universally pervaded to disassociate them into their elements. The disassociated gases thus sucked in at the solar poles at an extremely low temperature

THE PROBLEM OF SOLAR ENERGY. 23

are brought into a state of combustion by friction and condensation, thus generating new supplies of light and heat, and the gases thus reunited by combustion are again projected into space, to be again slowly disassociated by the operation of the sun's light and heat. The result of this combustion is to form aqueous vapor and carbonic acid and carbonic oxide, and these gases, when disassociated in space, are resolved into carbon, oxygen, and hydrogen, which again and again are thus recombined and again and again decomposed as they pass over the sun's surface (7).

The seven hypotheses above described are the only ones now in vogue, and a brief analysis will show that no single one of them, nor all combined, will give sufficient results to account for the essential difficulties or known conditions of the problem. The first and second hypotheses are answered by the fact set forth by Helmholtz (Popular Scientific Lectures, article "On the Origin of the Planetary System"), that, if the mass of the sun were composed of the two elements capable by combination of producing the greatest possible light and heat,— to wit, hydrogen and oxygen in the proportions in which they unite to form water,—" calculation shows that under the above supposition the heat resulting from their combustion would be sufficient to keep up the radiation of heat from the sun three thousand and twenty-one years. That, it is true, is a long time, but even profane history teaches that the sun has lighted and warmed us for three thousand years, and geology puts it beyond doubt that this period must be extended to millions of years."

The third hypothesis relates to *currents* of electricity. We have no knowledge of currents of electricity which could produce, however multiplied or intensified, such light and heat as are constantly poured forth from the sun into all space. That electricity is the intermediate cause of our sun's energy, and of all solar energy, it is the purpose of this work to demonstrate, but not electric *currents*, which find their attractiveness to theorists in the vague suggestion of which Professor Proctor speaks, referring to comets, in his article on "Cometic Mysteries," "that perhaps *this* is an electrical phenomenon; perhaps *that other feature* is electrical, too; perhaps *all or most* of the phenomena of comets depend on electricity." But he adds, "It is so easy to make such suggestions, so difficult to obtain evidence in their favor having the slightest scientific value. Still, I hold the electrical idea to be well worth careful study. Whatever credit may hereafter be given to any electrical theory of comets will be solely and entirely due to those who may help to establish it upon a basis of sound evidence,—none whatever to the mere suggestion, which has been made time and again since it was first advanced by Fontanelle." It will be seen that the present work, in demonstrating the true source and mode of solar energy, in itself presents a full and sufficient explanation of all the cometic mysteries referred to, as well as all those pertaining to other solar systems in space, and the multifarious phenomena which they present. Indeed, the philosophic mind will not be satisfied with the sufficiency of any hypothesis which will not unlock the mysteries and clearly explain the phenomena of other systems,—of comets, variable and temporary stars, double stars, and all the complicated celestial economy which to the eye of the mere observer presents a bewildering scene of the operation of independent and inscrutable forces. The fifth hypothesis cited, that of meteoric impact, doubtless plays a part, as we know from the generation of light and heat by the constant passage of similar bodies through our own atmosphere. And we know, of course, that the sun, by its vastly-increased attraction, must be subjected to the constant impact of such meteoric bodies in enormous numbers. But the fatal defect in the theory is that such impacts, to produce the radiant energy of the sun, must constantly add to its mass in like proportion, and as the motions and distances of the planets in their orbits are regulated and preserved by virtue of the substantially constant mass of the sun, any progressive and considerable increase in its mass must constantly bring the planets nearer and nearer, and thus increase their orbital velocity. Helmholtz quotes from Sir William Thomson's investigation, that, "assuming it to hold, the mass of the sun should increase so rapidly that the consequences would have shown themselves in the accelerated motion of the planets. The entire loss of heat from the sun cannot, at all events, be produced in this way; at the most a portion, which, however, may not be inconsiderable." R. Kalley Miller, in "The Romance of Astronomy," says, "But more recent observations have led Sir William Thomson to a modification of his theory. He has calculated that if the meteoric shower were sufficiently heavy to make up for the sun's whole expenditure of heat, the matter of the corona must be so dense as seriously to perturb the orbits of certain comets which pass very close to his surface,—a result which is found not to be the case. But the meteoric theory is only thrown back a step. If the sun's mass were originally formed, as is not at all improbable, by the agglomeration of these particles, Sir William Thomson has calculated that the heat generated by their thus falling together would be sufficient to account for a supply of twenty million years of solar heat at the present rate of emission. And thus, though the meteors are not sufficient to maintain the energy of our system unimpaired, they may yet have been the original storehouse from which all that energy was derived. . . . But if the economy of our system be spared long enough, the day must come when the sun with age has become wan; when the matter of the corona has all been drawn in and used up without avail; when the lavish luxuriance with which he has showered abroad his light and heat has finally exhausted all his stores. He has still power, aided by the resisting medium, to drag his satellites one by one down upon his surface; and the shock of each successive impact will, for a brief period, give him a fresh tenure of life. When the earth crashes into the sun it will supply him with a store of heat for nearly a century, while Jupiter's large mass will extend the period by nearly thirty thousand years. But when the last of the planets is swallowed up, the sun's energies will rapidly die out and a deep and deathly gloom gather about nature's grave. Looking into the ages of a future eternity, we can see nothing but a cold and burnt-out mass remaining of that glorious orb which went forth in the morning of time, joyful as a bridegroom from his chamber, and rejoicing as a strong man to run a race."

The sixth hypothesis is that to which most credence is now given. It is that of evolution of energy by condensation of volume. Professor Proctor ("The Sun as a Perpetual Machine") says, "In company with this great mystery of seeming waste comes the yet more difficult problem, how to explain the apparent continuance of solar light and heat during millions of years. We know from the results of geological research that the earth has been exposed to the action of the solar rays with their present activity during at least a hundred million years. Yet it is difficult to see how, on any hypothesis of the generation of solar heat, or by combining together all possible modes of heat generation, a supply for more than twenty millions of years in the past and a possible supply for as long a period in the future can be accounted for." Of these vast periods of terrestrial existence in the past we quote the following from a recent publication:

"Professor C. D. Wolcott expresses the opinion that geologic time is not to be measured by hundreds of years, but simply by tens of millions. This is widely different from the conclusion arrived at by Sir Charles Lyell, who, basing his estimate on modifications of certain specimens of marine life, assigned 240,000,000 years as the required geological period; Darwin claimed 200,000,000 years; Crowell, about 72,000,000; Geike, from 73,000,000 upward; McGee, Upham, and other recent authorities claim from 100,000,000 up to 680,000,000."

Helmholtz ("On the Origin of the Planetary System") says, "It is probable rather that a great part of this heat, which was produced by condensation, began to radiate into space before this condensation was complete. But the heat which the sun could have previously developed by its condensation would have been sufficient to cover its present expenditure for not less than 22,000,000 of years of the past. . . . We may therefore assume with great probability that the sun will still continue in its condensation, even if it only attained the density of the earth, though it will probably become far denser in its interior, owing to its far greater pressure; this would develop fresh quantities of heat, which would be sufficient to maintain for an additional 17,000,000 of years the same intensity of sunshine as that which is now the source of all terrestrial life." Of this process of condensation Professor Ball, in his recent work, "In the High Heavens," says, "It goes without saying that the welfare of the human race is necessarily connected with the continuance of the sun's beneficent action. We have indeed shown that the few other direct or indirect sources of heat which might conceivably be relied upon are in the very nature of things devoid of necessary permanence. It becomes, therefore, of the utmost interest to inquire whether the sun's heat can be calculated on indefinitely. Here is indeed a subject which is literally of the most vital importance, so far as organic life is concerned. If the sun shall ever cease to shine, then it must be certain that there is a term beyond which human existence, or indeed organic existence of any type whatever, cannot any longer endure on the earth. We may say once for all that the sun contains just a certain number of units of heat, actual or potential, and that he is at the present moment shedding that heat around with the most appalling extravagance." Quoting from Professor Langley, he says, "We feel certain that the incessant radiation from the sun must be producing a profound effect on its stores of energy. The only way of reconciling this with the total absence of evidence of the expected changes is to be found in the supposition that such is the mighty mass of the sun, such the prodigious supply of heat or what is the equivalent of heat which it contains, that the grand transformation through which it is passing proceeds at a rate so slow that, during the ages accessible to our observations, the results achieved have been imperceptible. . . . We cannot, however, attribute to the sun any miraculous power of generating heat. That great body cannot disobey those laws which we have learned from experiments in our laboratories. Of course no one now doubts that the great law of the conservation of energy holds good. We do not in the least believe that because the sun's heat is radiated away in such profusion it is therefore entirely lost. It travels off, no doubt, to the depths of space, and *as to what may become of it there we have no information.* Everything we know points to the law that energy is as indestructible as matter itself. The heat scattered from the sun exists at least as *ethereal vibration, if in no other form.* But it is most assuredly true that this energy, so copiously dispensed, is lost to our solar system. There is no form in which it is returned, or in which it can be returned. The energy of the system is as surely declining as the store of energy of the clock declines according as the weight runs down. In the clock, however, the energy is restored by winding up the weight, but there is no analogous process known in our system." The purpose of the present work, however, is to clearly demonstrate that just such a process is actually being carried on, and has been so carried on from the beginning, and will be forever. This writer continues reviewing the suppositions formerly entertained, that the sun was a heated body gradually cooling down, or that it was undergoing absolute combustion, and shows that they were utterly insufficient. He then refers to the theory of meteoric supply, of which he says, "It can, however, be shown that there are not enough meteors in existence to supply a sufficient quantity of heat to the sun to compensate the loss by radiation. The indraught of meteoric matter may, indeed, certainly tend in some small degree to retard the ultimate cooling of the great luminary, but its effect is so small that we can quite afford to overlook it from the point of view that we are taking in these pages. It is to Helmholtz we are indebted for the true solution of the long-vexed problem. He has demonstrated in the clearest manner where the source of the sun's heat lies. . . . A gaseous globe like the sun, when it parts with its heat, observes laws of a very different type from those which a cooling solid follows. As the heat disappears by radiation the body contracts; the gaseous object, however, decreases in general much more than a solid body would do for the same loss of heat. . . . The globe of gas unquestionably radiates heat and loses it, and the globe, in consequence of that loss, shrinks to a smaller size. . . . In the facts just mentioned we have an explanation of the sustained heat of the sun. Of course we cannot assume that in our calculations the sun is to be treated as if it were gaseous throughout its entire mass, but it approximates so largely to the gaseous state in the greater part of its bulk that we can feel no hesitation in adopting the belief that the true cause has been found."

Regarding the constitution of the sun, it may be stated, however, that we only see its photosphere, which is the visible sun, and the whole volume has a density about that of water; but no man has ever seen the body of the sun itself. In this respect it is like the planet Jupiter: we only know that its density cannot be less than one-fourth the density of the earth's solid globe. If the photosphere extend to a depth of one thousand, ten thousand, or a hundred thousand miles, the density of the sun's body or core will be correspondingly increased. Even computing the whole visible volume, the density is far greater than that of any gas we know, even with the solar pressure of gravity; with the sun's metallic vapors, if the whole core were already vaporized, we would not, to say the least, be likely to observe the sun-spots and other solar phenomena as we find them actually to occur; this, however, will be more fully considered later on. The author continues, "But there is a boundary to the prospect of the continuance of the sun's radiation. Of course, as the loss of heat goes on the gaseous parts will turn into liquids, and as the process is still further protracted the liquids will transform into solids. Thus, we look forward to a time when the radiation of the sun can be no longer carried on in conformity with the laws which dictate the loss of heat from a gaseous body. When this state is reached the sun may, no doubt, be an incandescent solid with a brilliance as great as is compatible with that condition, but the further loss of heat will then involve loss of temperature. . . . There seems no escape from the conclusion that the continuous loss of solar heat must still go on, so that the sun will pass through the various stages of brilliant incandescence, of glowing redness, of dull redness, until it ultimately becomes a dark and non-luminous star. . . . There is thus a distinct limit to man's existence on the earth, dictated by the ultimate exhaustion of the sun. . . . The utmost amount of heat that it would ever have been possible for the sun to contain would, according to this authority (Professor Langley), supply its radiation for eighteen million years at the present rate. . . . It seems that the sun has already dissipated about four-fifths of the energy with which it may have originally been endowed. At all events, it seems that, radiating energy at its present rate, the sun may hold out for four million years or for five million years, but not for ten million years. . . . We have seen that it does not seem possible for any other source of heat to be available for replenishing the waning stores of the luminary." He concludes by saying that the original heat may have been imparted as the result of some great collision, the solar body having itself been dark before the collision occurred, and that it may be reinvigorated by a repetition of a similar startling process, but indicates in general terms that such an operation would be bad for the round world and all contained therein. It would, in fact, be rough treatment for even a hopeless case.

Condensation of the solar volume is unquestionably a source of heat, for we know that the solid or liquid interior of the earth increases in temperature at a definite ratio as we descend through its crust; but long before the sun shall have become contracted to the density of the earth all its heat will have become substantially internal heat, and it can then supply no more by radiation to its surrounding planets.

It will be seen that the radiant energy of the sun on any of the above hypotheses is not sufficient to account even for the life period of the earth in the past, and that its future period of energy must be still more brief. Professor Ball ("In the High Heavens"), basing his views on Laplace's "Nebular Hypothesis," says, "Looking back into the remote ages, we thus see that the sun was larger and larger the further back we project our view. If we go sufficiently far back, we seem to come to a time when the sun, in a more or less completely gaseous state, filled up the surrounding space out to the orbit of Mercury, or, earlier still, out to the orbit of the remotest planet." According to this hypothesis, all these brilliant suns, the author says, will "settle down into dark bodies like the earth," and that "every analogy would teach us that the dark and non-luminous bodies in the universe are far more numerous than the brilliant suns. We can never see the dark objects; we can discern their presence only indirectly. All the stars that we can see are merely those bodies which at this epoch of their career happen for the time to be so highly heated as to be luminous. . . . It may happen that there are dark bodies in the vicinity of some of the bright stars to which these stars act as illuminants, just in the same way as the sun disperses light to the planets." One would naturally suppose, however, that there must be some sort of laws to govern such stupendous operations, and that nature is not merely engaged in blowing bubbles. To quote Professor Newcomb:

THE PROBLEM OF SOLAR ENERGY. 35

"At the present time we can only say that the nebular hypothesis is indicated by the general tendencies of the laws of nature; that it has not been proved to be inconsistent with any fact; that it is *almost a necessary consequence of the only theory by which we can account for the origin and conservation of the sun's heat;* but that it rests on the assumption that this conservation is to be explained by the laws of nature as we now see them in operation. Should any one be sceptical as to the sufficiency of these laws to account for the present state of things, science can furnish no evidence strong enough to overthrow his doubts until the sun shall be found growing smaller by actual measurement, or the nebulæ be actually seen to condense into stars and systems."

While the validity of the views set forth in the present volume does not depend on the sufficiency or insufficiency of the nebular hypothesis, and in fact requires the condensation as well as the expansion of the solar volume *under the influence of heat* to be recognized and its extreme importance pointed out, yet it must not be supposed that this great generalization of Kant and Laplace, based on the views presented originally by Sir William Herschel, is established, or that the difficulties in its way are not so enormous as to be almost insuperable. Professor Ball points out that thousands of bodies occupy our solar system, and together compose it as a whole; that these have orbits of every sort of eccentricity and direction, and occupying all possible planes which can pass through the sun;

that the bodies circle around the sun, some backward and others forward, and that only the planets seem to conform to some common order; and without this order, which may be accidental, so far as our knowledge goes, the system would have been disrupted long since, if it ever could have begun its operations; and that in this view the heavens may be strewn with wrecks of systems which failed to survive from inherent want of harmony,—that is to say, as based on observation only. Whether the nebular hypothesis be a universal or a partial law of development, or whether the real processes be quite different, cannot, however, depend on the continued maintenance and evolution of the sun's energy, as this source must in truth be sought for in quite a different direction.

The remaining hypothesis (the seventh) is considered in detail in Professor Proctor's work, "Mysteries of Time and Space." The fatal defect in Dr. Siemens's theory is, that his gases will not be projected from the sun's equator. Professor Proctor says, "Thus the centripetal tendency of matter at the sun's equator is very much greater (many hundreds of times greater) than its centrifugal tendency, and there is not the slightest possibility of matter being projected into space from the sun's surface by centrifugal tendency. Nor is there any part of the sun's mass where the centrifugal tendency is greater than at the surface near the equator. So that, whatever else the sun may be doing to utilize his mighty energies, he is certainly not throwing off matter constantly from his equatorial regions, as Dr. Siemens's theory requires." There are other difficulties which Professor Proctor considers, such as the doubt as to the power of the sun's rays to disassociate combined gases in space, and also that, since both light and heat must be utilized in this work, if the sun's energies are to be perpetually renewed, these forces would sensibly disappear in work, and the result would be that the fixed stars would be invisible beyond their domains, and their light, when not totally cut off, would be greatly diminished, in any event, as distances increased, which is not the case. Besides, these gases thus disassociated could never be entirely used by the sun, and the remainder would be wasted, and the part wasted would vastly exceed that utilized, probably in as great proportion of waste as that of the sun's light not utilized by the planets, which gather but one two-hundred-and-thirty-two-millionths of the whole. It may be further added that these gases would be mechanically mixed, the combined and the disassociated, and this would be mostly the case in those parts nearest the sun, so that large volumes of spent and useless gases would have to be carried in to no purpose whatever. In fact, these gases would gradually form a closed circuit of supply and discharge, and surrounding space would be but slightly affected. Professor Proctor concludes, " We have, in fact, the fallacy of perpetual motion in a modified form."

It will be apparent that under any single one, or all, of these hypotheses, the future prospect for created forms and continued existence is hopeless, and that the inevitable result must do violence to every conception of either an intelligent creative power or the operations of universal law. The mind revolts from the continued degradation and destruction of all organic creation, while the malevolent and iconoclastic forces of nature hold high revel over final ruin and eternal destruction, brought about by their own incessant efforts, striking out blindly to make or mar, and they alone the deathless survivors, the half-blind fates and furies of the eternal future. It betokens, not the processes of orderly government, but the reign of anarchy.

NOTE.—Since this work has been in press, at the annual meeting of the British Association, August 8, 1894, Lord Salisbury, the President, delivered a powerful and lucid address on the present status of scientific knowledge and its limitations. With reference to the antiquity of the earth we quote the following: "It is evident, from the increase of heat as we descend into the earth, that the earth is cooling, and we know, by experiment within certain wide limits, the rate at which its substances—the matters of which it is constituted—are found to cool. It follows that we can approximately calculate how hot it was so many million years ago; but if at any time it was hotter at the surface by fifty degrees Fahrenheit than it is now, life would then have been impossible upon the planet, and, therefore, we can without much difficulty fix a date before which organic life on earth cannot have existed. Basing himself on these considerations, Lord Kelvin limited the period of organic life upon the earth to a hundred million years, and Professor Tait, in a still more penurious spirit, cut that hundred down to ten." If a period of anything like ten million years, even, has been requisite to cool the earth's surface only fifty degrees in temperature, what time must have elapsed since the terrestrial globe had a temperature high enough to effect the difficult chemical combinations of many of the elements which compose its structure? And even this must have been far less than the vast cycles of time during which original consolidation was effected. Through all these ages the sun must have been pouring out his radiant energy at at least his present rate. Radiation of heat from the earth may have been relatively less rapid from a denser carbon-laden atmosphere in times past than at present, but it never could have been more so. The whole address cited is, indeed, strongly corroborative of the facts upon which the present work is based.

CHAPTER II.

THE CONSTITUTION AND PHENOMENA OF THE SUN.

THE various theories thus reviewed, while not sufficient in themselves to account for the facts of our own solar system, are fatally defective in another respect. While they aim to account for the sun's light and heat, they all fail to consider the active medium of the solar light and heat in the sun itself. It is not simply a highly-heated central mass glowing in space. It is a vast orb surrounded by different envelopes of incandescent vapors or gases, and by far the most vast in volume, as well as in light and heat-radiating power, are the photosphere and its superincumbent chromosphere, composed almost entirely of free hydrogen gas in a state of intense incandescence. Whence comes this enormous mass of hydrogen? And how explain the entire absence of free hydrogen gas from our own atmosphere and its replacement by oxygen? There is a recent theory propounded by Mr. A. Mott, which is set forth in detail in Professor Ball's "In the High Heavens," and which endeavors to account for the remarkable absence of free hydrogen gas from the earth's atmosphere, for, as the author states, "It is a singular fact that hydrogen in the free state is absent from our atmosphere." The theory, in brief, is that the molecules of hydrogen gas have an average speed of about a mile a second,—which, however, is only one-seventh that required to shoot them off into space,—but that these molecules are continually changing their velocity, and may sometimes attain a speed of seven miles a second; the result is that " every now and then a molecule of hydrogen succeeds in bolting away from the earth altogether and escaping into open space." During past ages the molecules of hydrogen would thus have gradually wiggled up through the air, and finally disappeared into outer darkness for good and all; and thus " the fact that there is at present no free hydrogen in the air over our heads may be accounted for." Since the molecules of oxygen have only a velocity of a quarter mile a second, that unfortunate gas remains behind and is consumed.

The first difficulty with this theory is to explain how, if the hydrogen wiggled off in this unceremonious manner, it ever wiggled on. There is no objection to a gait of this rapidity, however; it is highly creditable, in fact; but we have a right to expect some degree of consistency in even so light-headed a body as hydrogen gas. The article quoted thus continues: " If the mass of the earth were very much larger than it is, then the velocities with which the molecules of hydrogen wend their way would never be sufficiently high to enable them to quit the earth altogether, and consequently we might in such a case expect to find our atmosphere largely charged with hydrogen." It will be seen that, according to this theory, hydrogen is able to achieve a speed of seven miles per second under exceptional excitement, and that this molecular velocity is just enough, and no more than enough, to give it egress. We know that Jupiter's mass is three hundred times as great as that of the earth, and the attraction of gravity is so powerful on the surface of that planet that, as the writer just quoted says, "Walking, or even standing, would involve the most fearful exertion, while rising from bed in the morning would be a difficult, indeed, probably, an impossible, process." We also know that the atmosphere of this planet is laden with enormous clouds floating at various altitudes and with incessant movements. We are told that "the molecular speed of aqueous vapor averages only one-third of that attained by the molecules of hydrogen." Of course, on the planet Jupiter, hydrogen would have no chance of escape at all: it would just have to stay and take it, like the rest of us. Jupiter must thus have an atmosphere like our own, except that it is "largely charged with hydrogen." Of the clouds upon this planet, Professor Ball says, "In fact, the longer we look at Jupiter the more we become convinced that the surface of the planet is swathed with a mighty volume of clouds so dense and so impenetrable that our most powerful telescopes have never yet been able to pierce through them down to the solid surface of the planet." With the densities, molecular velocities, and specific gravity of the oxygen, nitrogen, and the hydrogen, with which latter the atmosphere of Jupiter must be "largely charged," as it is said, it is difficult to understand how such enormous clouds of aqueous vapors, themselves composed of oxygen, which is a very slow-footed gas, and hydrogen, could travel about with such facility; we ought to find them packed down like London fog, to say the least, upon the surface of that planet, with the supernatant gases all adrift overhead. Jupiter is a hot body; it has not yet cooled down; and if it is provided with volcanoes, such as its great red spot and the analogies of the earth and moon would suggest, we can tell pretty nearly what would have happened long ago with a Jovian atmosphere like ours; but "largely charged with hydrogen," if we compare it with, say, an equal mass of dynamite touched off by a volcanic explosion; there would not have been enough of old Jupiter left to swear by, and what was left would not have had any atmosphere at all. On Mars, the same writer thinks the oxygen would still cling, like the fragrance of the rose, but that all the molecules of the fleet-footed and excitable hydrogen would long since have taken French leave, as it did from the earth; but at the moon, on account of its small size and mass, both gases would have gone off incontinently together. "It is now easy," the author says, "to account for the absence of atmosphere from the moon. . . . Neither of the gases, oxygen or nitrogen, to say nothing of hydrogen, could possibly exist in the free state on a globe of the mass and dimensions of our satellite. . . . Indeed, the weight of every object on the moon would be reduced to the sixth part of that which the same object has on earth." Nevertheless, it may be said that the moon has considerable weight, as weights go, but with a comet it is quite a different matter. "These bodies," the author says, "demonstrate conclusively that the quantity of matter even in a comet is extremely small when compared with its bulk. The conclusion thus arrived at is confirmed by the fact that our efforts to obtain the weight of a comet have hitherto proved unsuccessful. . . . It has thus been demonstrated that, notwithstanding the stupendous bulk of a great comet, its mass must have been so inconsiderable as to have been insufficient to disturb even such unimportant members of the solar system as the satellites of Jupiter." Now, here is a state of things; for the spectroscope shows that comets are fully provided with a large supply of hydrogen, enough and to spare for ornament, even, and of nitrogen also, while it is the abnormally fugacious oxygen which has, apparently, taken its departure. Of course, such facts demonstrate the untenability of the theory, which is, besides, in direct contradiction with the laws governing gaseous diffusion. Gases pass into each other with the same velocity as into a vacuum, and it is not to be imagined that the molecules of hydrogen could thus move individually off, unless forced upward by the pressure of some other gas, which the law of gaseous diffusion makes impossible. We should as readily expect to see a tumbler full of iron balls, into the interstices of which loose sand has been poured, manifest a similar phenomenon by the wiggling out of the less dense sand at the top of the glass. One might also ask whence, if this theory had any substantial basis, could come the enormous volumes of hydrogen gas in the atmosphere of a new or temporary star, in a few hours, or the changes manifested in the atmospheres of the variable stars. So, also, the nebular or any other hypothesis of creation would be impossible under this theory, as the heavier and less mobile gaseous elements would remain behind, or be condensed nearest the center of gravity of the aggregating nebula, while the more rapid gases would disappear outwardly, and in consequence the sun would be found to be composed of the heavier elements exclusively, and each of the planets, in turn, would consist of only one or two elements, in accordance with the more and more mobile character of their molecular movements, and the uniformity of chemical constitution between the sun and planets, as well as the fixed stars, would not be found to exist. The theory, in fact, is an example of the endeavor to explain an easily understood difficulty by a less easily understood impossibility.

None of the different theories even attempt to account for the prodigious volumes of hydrogen in the solar atmosphere, and without its presence the sun, so far as we know, would be almost an inert mass, considered as a source of energy for the supply of our planetary system. We know, of course, that meteors contain sometimes as much as six volumes of gases, largely composed of hydrogen, at our own atmospheric pressure. But the pressure at the sun's surface is more than twenty-seven times that at the surface of the earth, and yet the volume of hydrogen there existing visibly is vaster beyond computation than any possible mass of meteoric material could supply. So, also, while it may be granted that condensation of volume must vastly raise the solar temperature, how could it produce the enormous masses of hydrogen, the lightest of all the elements, unless they have been temporarily occluded and finally thrown out from within, which is impossible? These vast volumes of hydrogen are to be considered first of all in any attempt whatever to solve the problem of the source and mode of solar energy.

Considering the phenomena presented within the limits of our own solar system alone, we find that the earth is one of a single family of planets, each of which very closely resembles it, and all of which circle, in slightly elliptical orbits, at various distances around the sun, their orbits occupying substantially the same plane, thus making our solar system a flat disk of space occupied by the sun as a center, with the planets and their satellites moving harmoniously around it. The planets differ from each other in size, mass, and temperature, but each is surrounded by an envelope of aqueous vapor, suspended in an atmosphere substantially like our own. Professor Proctor, in his "Light Science for Leisure Hours," says of the planet Jupiter, "His real surface is always veiled by his dense and vapor-laden atmosphere. Saturn, Venus, and Mercury are similarly circumstanced." Of Mars he says that it is "distinctly marked (in telescopes of sufficient power) with continents and oceans which are rarely concealed by vapors." Now, whence comes this aqueous vapor surrounding all the planets? Whether received originally from the diffused nebular mass from which our solar system is supposed to have been condensed, or attracted by the force of gravity from interplanetary space, like the meteors which fall upon the earth's surface, it is evident that interplanetary space must once have been pervaded with aqueous vapor, since the nebular mass from which our solar system was constituted must have occupied at least the space embraced within its largest planetary orbit, and doubtless much more; and if so, such aqueous vapor, and other vapors also, must still persist in space, just as the meteoric particles which so constantly manifest themselves in our atmosphere. If the planets had no common origin, the evidence is equally conclusive, since then this identical substance could only have been derived from a common source, which can only be interplanetary space. This also is in accordance with the laws of attraction, which would operate to gather and condense the rarefied aqueous vapor of space around the planetary masses in definite proportions. In his "Familiar Essays on Scientific Subjects," Professor Proctor says, "In fact, we do thus recognize in the spectra of Mars, Venus, and other planets the presence of aqueous vapor in their atmosphere;" and in his "Mysteries of Time and Space" he says, "We may admit the possibility that the aqueous vapor and carbon compounds are present in stellar or interplanetary space." But in addition to this aqueous vapor which surrounds the planetary bodies, we find free oxygen in vast quantities, and, with this, free nitrogen in mechanical admixture, and these together constitute the atmosphere we breathe, and which sustains organic life by a process of slow combustion. But we find no free hydrogen either in our own atmosphere or in that of other planets. Turning now to the sun, we find that it is surrounded by an atmosphere as well as the planets, but that this atmosphere is composed not of free oxygen, but of free hydrogen. In his article, "Oxygen in the Sun," Professor Proctor says, "Fourteen only of the elements known to us, or less than a quarter of the total number, were thus found to be present in the sun's constitution; and of these all were metals, if we regard hydrogen as metallic. . . . But most remarkable of all, and most perplexing, was the absence of all trace of oxygen and nitrogen, two gases which could not be supposed wanting in the substance of the great ruling center of the planetary system." The researches of Dr. Draper indicated, however, that oxygen could be found in the sun; not in his external atmosphere but far down within his surface. Professor Proctor says, "Dr. Draper mentions that he has found no traces of oxygen above the photosphere." Such free oxygen cannot be associated with the hydrogen, however, even if its presence be finally determined, but it may be due to the deoxidation of solid compounds precipitated upon the sun from space, and held at a temperature above that of disassociation, as hydrogen is sometimes generated at the surface of the earth.

The vast mass of the solar atmosphere is composed of hydrogen gas, with which are found commingled vapors of the various elements which enter into the sun's constitution, and this solar atmosphere corresponds in proportion, speaking generally, with our own atmosphere, except that the volume of solar hydrogen is vastly greater than that of terrestrial oxygen, for the reason, as will be explained, that water contains two volumes of the former to one of the latter.

In Appleton's Cyclopædia the sun is thus described, (article by Professors Langley and Proctor): "To sum up briefly the received hypotheses of the physical constitution of the sun: of its internal structure we know nothing, but we can infer, from the low density of the solar globe as a whole, that no considerable portion is solid or liquid. The regions we examine appear to consist of cloud layers at several levels floating in a complex atmosphere, in which probably most of the elements are known to us, and certainly many of them exist in the form of vapor. Outside this complex atmosphere extend envelopes of simpler constitution, though into them occasionally arise the vapors which ordinarily lie lower down. The sierra, for instance, consists in the main of glowing hydrogen gas and that gas, whatever it may be, which produces the line near the orange-yellow sodium lines. The prominence region may be regarded as simply the extension of the sierra." Of these prominences,

Professor Ball says, "The memorable discovery made by Janssen and Lockyer, independently, in 1868, showed that the prominences could be observed without the help of an eclipse, by the happy employment of the peculiar refrangibility of the rosy light which these prominences emit. . . . We can now obtain, not, as heretofore, merely isolated views of special prominences through the widely opened slit of the spectroscope, but we are furnished, after a couple of minutes' exposure, with a complete photograph of the prominences surrounding the sun. . . . The incandescent region of the chromosphere from which these prominences arise is also recorded with accuracy." Resuming our quotation from Appleton's Cyclopædia: "The inner corona is still simpler than the sierra, so far as its gaseous constitution is concerned; but here meteoric and cometic matter appears, extending to the outer corona and to great distances beyond even the visible limits of the zodiacal. Returning to the photosphere, we find it subject to continual fluctuations, both from local causes of agitation and from the subjacent vapor acting by its elasticity to burst through it; the faculæ, which are found to be above the general level of the photosphere, are taken to be heapings up of the luminous matter like the crested surges of the sea. All the strata are subject to great movements, which sometimes have the character of uniform progression analogous to our trade-winds, and sometimes are violent, and resemble in their effects our tornadoes and whirlwinds. Eruptive action appears to operate from time to time with exceeding violence, but whether the enormous velocities of outrush are due to true explosive action (which would compel us to believe that the sun is enclosed by a liquid shell, so as to resemble a gigantic bubble) or to the uprising of lighter vapors from enormous depths, as heated currents rise in our own atmosphere, is not as yet certainly known." The sierra, or chromosphere, is thus described in the same article: "The sierra presents four aspects: 1, smooth with defined outline; 2, smooth but with no defined outline; 3, fringed with filaments; and, 4, irregularly fringed with small flames. The prominences may be divided into three orders,—heaps, jets, and plumes. The heaped prominences need no special description. The jets . . . originate generally in rectilinear jets either vertical or oblique, very bright and very well defined. They rise to a great height, often to a height of at least eighty thousand miles, and occasionally to more than twice that; then bending back, fall again upon the sun like the jets of our fountains. Then they spread into figures resembling gigantic trees more or less rich in branches. Their luminosity is intense, insomuch that they can be seen through the light clouds into which the sierra breaks up. Their spectrum indicates the presence of many elements besides hydrogen. When they have reached a certain height they cease to grow, and become transformed into exceedingly bright masses, which eventually separate into fleecy clouds. The jet prominences last but a short time—rarely an hour, frequently but a few minutes,—and they are

CONSTITUTION AND PHENOMENA OF THE SUN. 51

only to be seen in the neighborhood of the spots. Wherever there are jet prominences there also are faculæ. The plume prominences are distinguished from the jets in not being characterized by any signs of an eruptive origin. They often extend to an enormous height; they last longer than the jets, though subject to rapid changes of figure; and, lastly, they are distributed indifferently over the sun's surface. It would seem that in the jets a part of the photosphere is lifted up, whereas in the case of plumes only the sierra is disturbed." Of these eruptions Professor Ball says, "Vast masses of vapors are frequently expelled from the interior of the sun by convulsive throes with a speed of three hundred, four hundred, and sometimes nearly a thousand miles a second. . . . The spectroscope enables the observer actually to witness the ascent of these solar prominences."

The corona, which extends beyond the chromosphere, has been determined by its continuous spectrum to be a vast envelope extending at least a million miles from the sun's surface. "It cannot be a solar atmosphere," Professor Proctor observes in his article on this subject, in his "Mysteries of Time and Space.". . . "It will be seen, then, how inconceivably great the pressure exerted by a solar atmosphere some eight thousand times as deep as ours would necessarily be, let the nature of the gases composing it be what it may.". . . "If a man could be placed on the solar surface, his own weight would crush him as effectually as though while on earth a weight of a couple of tons were heaped upon him. . . . Now, it happens that we know quite well that the pressure exerted by the real solar atmosphere, even close by the bright surface which forms the visible globe of the sun, is nothing like so great as it would be if the corona formed part of that atmosphere." In the article "Sun," in Appleton's Cyclopædia, it is stated that "Mr. Arthur W. Wright, of Yale College, has succeeded in showing that this light (the zodiacal) is not emitted from incandescent gas, but reflected from particles or small bodies, and hence derived from the sun.". . . " There is reason to believe that the true solar corona extends much farther (than a million miles), and that, in reality, the zodiacal light forms the outer part of the solar corona." Proctor, again, in his article on the corona, says, " It would seem to follow that the corona is due to bodies of some sort travelling around the sun, and by their motion preserved either from falling towards him (in which case the corona would quickly disappear) or from producing any pressure upon his surface, as an atmosphere would." In his article on " The Sun as a Perpetual Machine," he says, " There is every reason for regarding the zodiacal as consisting in the main of meteorolithic masses, a sort of cosmical dust, rushing through interplanetary space with planetary velocities. To such matter, assuming, as we well may, that space really is occupied by attenuated vapors, . . . the luminosity of the zodiacal would be attributable to particles of dust emitting light reflected by the sun or by phosphorescence (this last may be seriously

CONSTITUTION AND PHENOMENA OF THE SUN. 53

questioned). But there is another cause for luminosity of these particles which may deserve a passing consideration. Each particle would be electrified by gaseous friction in its acceleration, and its electric tension would be vastly increased in its forcible removal, in the same way as the fine dust of the desert has been observed by Werner Siemens to be in a state of high electrification on the apex of the Cheops Pyramid. Would not the zodiacal light also find explanation by slow electric discharges backward from the dust towards the sun?" It may be observed in passing that such electrical glow is much more prominently, and more likely to be, the result of induction than of friction. In the article "Sun," previously quoted, Professor Young says, "There is surrounding the sun, beyond any further reasonable doubt, a mass of self-luminous gaseous matter, whose spectrum is characterized by the green line 1474 Kirchhoff. The precise extent of this it is hardly possible to consider as determined, but it must be many times the thickness of the red hydrogen portion of the sierra, perhaps, on an average, 8' or 10', with occasional horns of twice that height. It is not at all unlikely that it may even turn out to have no upper limit, but to extend from the sun indefinitely into space." In the same article the sun's apparent diameter is placed at about 32', so that the thickness of the above gaseous envelope would be not less than one-fourth the sun's diameter, or more than two hundred thousand miles. This coronal envelope, extending out from the solar body until gradually merged into the attenuated matter of space, has a light so feeble that it can only be clearly observed during total eclipse. Professor Ball (" In the High Heavens") says, " The sunlight is so intense that if it be reduced sufficiently by any artifice, the coronal light also suffers so much abatement that, owing to its initial feebleness, it ceases altogether to be visible." During the great eclipse of 1893 it was photographed, and of these photographs the same author says, " One of the most remarkable features in the structure of the corona is the presence of streamers or luminous rays extending from the north and south poles of the sun. *These rays are generally more or less curved*, and it is doubtful whether the phenomena they exhibit are not in some way a consequence of the rotation of the sun. This consideration is connected with the question as to how far the corona itself shares in that rotation of the sun with which astronomers are familiar. I should perhaps rather have said that rotation of the sun's photosphere which, as the sun-spots prove, is accomplished once every twenty-five days. Even this shell of luminous matter does not revolve as a rigid mass would do. By some mysterious law the equatorial portions accomplish their revolution in a shorter period than is required by those zones of the photosphere which lie nearer the north and south poles of the luminary. As to how the parts of the sun which are interior to the photosphere may revolve, we are quite ignorant. . . . We have no means of knowing to what extent the corona shares in the rotation. It would seem certain that the lower parts which lie comparatively near the surface must be affected by the rapid rotation of the photosphere; but it is very far from certain that this rotation can be shared to any great extent by those parts of the corona which lie at a distance from the sun's surface as great as the solar radius or diameter. . . . The corona presents a curious green line that seems to denote some invariable constituent of the sun's outer atmosphere, but the element to which this green line owes its origin is wholly unknown." The same author quotes from Dr. Huggins as follows: " It is interesting to read what Dr. Huggins has to tell us about the solar corona. The nature of this marvellous appendage to the sun is still a matter of uncertainty. There can, however, be no doubt that the corona consists of highly-attenuated matter *driven outward from the sun by some repulsive force*, and it is also clear that if this force be not electric, it must at least be something of a very kindred character. . . . So far as the spectrum of the corona is concerned, we may summarize what is known in the words of Dr. Huggins: 'The green coronal line has no known representative in terrestrial substances, nor has Schuster been able to recognize any of our elements in the other lines of the corona.'" The account given by General Myer—quoted in Professor Proctor's article, "The Sun's Corona"—of the great eclipse of 1869, as viewed from an altitude of five thousand five hundred feet above sea-level, is as follows: "As a centre stood the full and intensely black disk of the moon, surrounded by an aureola of soft bright light, through which shot out, as if from the circumference of the moon, straight, massive silvery rays, seeming distinct and separate from each other, to a distance of two or three diameters of the lunar disk; the whole spectacle showing as upon a background of diffused rose-colored light. The silvery rays were longest and most prominent at four points of the circumference, . . . apparently equidistant from each other. There was no motion of the rays: they seemed concentric." Three diameters would make these rays extend two and a half million miles at least from the sun's photosphere, or even its chromosphere. The coincidence between these rays and those observed (see above) in the eclipse of 1893 must be noted, since these latter were conceived at one time to be meteor streams. As those seen in 1893 radiated from the poles, and were curved in form, while those last noted radiated at four equidistant points, none polar, and were straight, it will be seen that, if both phenomena were of the same class, they could not have been due to meteor streams.

The sun's spots, which we will next refer to, are deep, relatively dark, but in fact extremely bright depressions in the photosphere. "Many spots are of enormous size" (see article, "Sun"); "one had a diameter exceeding fifty thousand miles, and many far larger than this have been seen. The spots are not scattered over the whole surface of the sun, but are for the most part confined to two belts between latitude five degrees and thirty degrees, on

CONSTITUTION AND PHENOMENA OF THE SUN. 57

either side of the solar equator. An equatorial zone six degrees wide is almost entirely free from

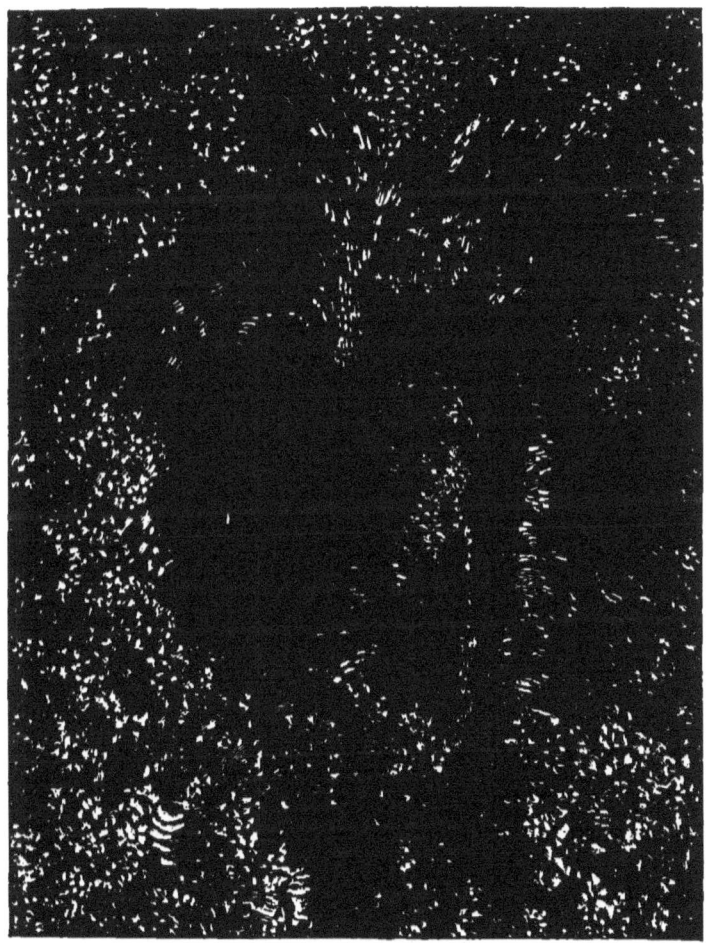

A typical sun-spot. (From the *Popular Science Monthly*, 1885.)

spots. . . . The inclination of the solar equator is about seven degrees. . . . The spots on the sun usually have a dark central region called the *umbra*, within which is a still darker part called the *nucleus*, while around this there is a fringe of fainter shade than the umbra, called the *penumbra*. Although the umbra and nucleus appear dark, however, it is not to be supposed that they are really dark; . . . though the nucleus looks perfectly black by contrast with the general surface, it shines in reality with a light unbearably brilliant when viewed alone, while his thermal measurements show that the heat from the nucleus is even greater proportionately than the light, and not very greatly below the heat of the surrounding surface. . . . The recognition of a nucleus within the umbra would seem to indicate that a third cloud layer (besides the outer or photosphere and a darker cloud layer beneath) exists within the second or internal layer of Herschel's theory. But the observations of Professor Langley show that most probably all the features of the solar photosphere yet observed are phenomena of cloud envelopes, since he has been able to recognize cloud forms at one level floating over cloud forms at a lower level, while even in the (relatively) darkest depths of the nucleus clouds are still to be perceived, though so deep down that their outlines can be barely discerned." Professor Ball says of the heat-wave of 1892, " As to the activity of the sun during the past summer, a very striking communication has recently been made by one of the most rising American astronomers, Mr. George E. Hale, of Chicago. He has invented an ingenious apparatus for photographing on the same plate at one exposure both the bright spots and the protuberances of the sun. . . . On the 15th of July a photograph of the sun showed a large spot. Another photograph taken in a few minutes exhibited a bright band; twenty-seven minutes later a further exposure displayed an outburst of brilliant faculæ all over the spot. At the end of an hour the faculæ had all vanished and the spot was restored to its original condition. It was not a mere coincidence that our magnetic observatories exhibited considerable disturbances the next day, and that brilliant auroras were noted." Carrington's observations have shown that spots in different solar latitudes travel at different rates. "Taking two parts of the visible solar surface in the same longitude, but one in latitude forty-five degrees (say), the other on the equator, the latter will advance farther and farther in longitude from the former, gaining daily about two degrees, so that in the course of about one hundred and eighty days it will have gained a complete revolution. That is to say, the sun's equator makes about two revolutions more per annum than regions in forty-five degrees north and south solar latitude." The sun is about 850,000 miles in diameter; its density is one-fourth that of the earth; its mass is 316,000 times greater, and its volume 1,253,000. Gravity at its surface is 27.1 times that of the earth; its distance is approximately 92,000,000 miles; it rotates upon its axis, which is inclined to the planetary plane at an angle of seven degrees, once in twenty-five and one-third days, apparently increased to thirty days by the earth's orbital advance in the same direction around the sun; and it has a motion around its center,—a true orbital motion,—due to displacement by gravity of the planetary masses, which, however, is always within its own mass.

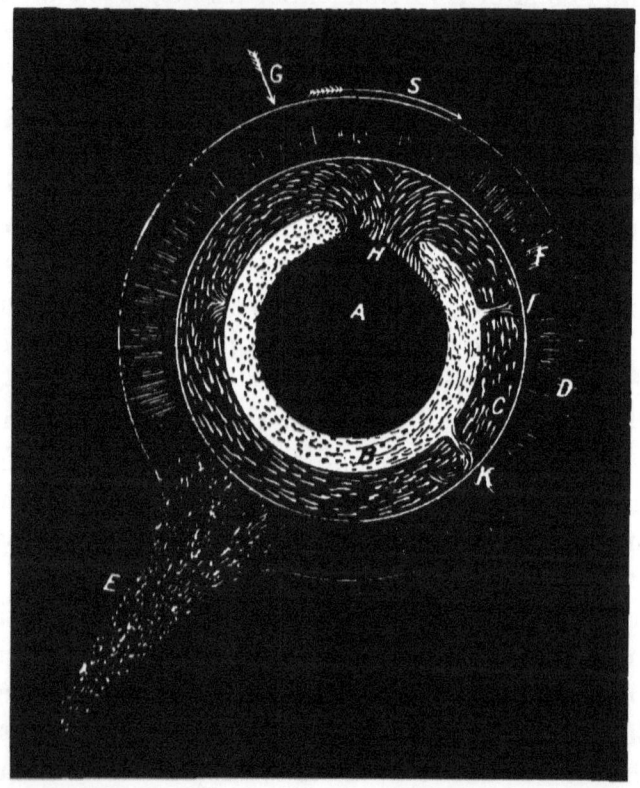

Structure of the sun.—A, solar core, or nucleus; B, photosphere, the visible orb; C, chromosphere, or sierra; D, corona, fading off into space; E, sun's long streamer; F, over faculæ in C and B; G, direction of line of planetary energy; H, active stage of a sun-spot; I, plume prominence; K, jet prominence; S, direction of sun's rotation.

The above, in brief, is, so far as we know, the constitution of the sun and its appendages. Its internal globe is surrounded by a glowing gaseous

CONSTITUTION AND PHENOMENA OF THE SUN. 61

envelope, the photosphere, which is the visible orb, composed of cloud masses of glowing hydrogen gas intermingled with vapors of many of our terrestrial elements, all in a state of apparent disassociation. Of the constitution of the sun's mass, Professor Ball says, "Professor Rowland has shown that thirty-six terrestrial elements are certainly indicated in the solar spectrum, while eight others are doubtful. Fifteen elements have not been found, though sought for, and ten elements have not yet been compared with the sun's spectrum. Reasons are also given for showing that, though fifteen elements had no lines corresponding to those shown in the solar spectrum, yet there is but little evidence to show that they are really absent from the sun. Dr. Huggins epitomizes these very interesting results in the striking remark, 'It follows that if the whole earth were heated to the temperature of the sun, its spectrum would resemble very closely the solar spectrum.'" Outside the photosphere is the simpler chromosphere, composed largely of hydrogen, and merging into the corona at a distance of hundreds of thousands of miles from the sun's apparent surface, and this corona extends outward to a vast distance, and is itself largely composed of self-luminous matter, the action of gravity being counterbalanced by the centrifugal force of orbital rotation, or more probably by electrical repulsion. The metallic vapors in the sun's photosphere are suspended in glowing hydrogen, which vastly preponderates over all the others in mass and volume, the incandescence of which is the principal source of solar light and heat. The planets revolve in elliptical orbits around this central sun, and crossing these orbits at various angles rush streams of cometic matter and comets and meteoric bodies, in streams and clouds, which, swiftly sweeping around at various distances, are again thrown off into space. Meteors constantly fall into the sun's mass, as they do upon the earth; but the grand key-note of all his life and energy, so far as we can perceive, is the vast envelope of glowing hydrogen gas.

Conversely, the planetary envelopes are of relatively cool oxygen mixed with nitrogen gas, which hold in suspension diffused aqueous vapors. If our own aqueous vapors are derived by the attraction of gravity from the interplanetary space, as they must have been, we can be sure that, were the sun at a sufficiently low temperature, he, too, would gather to himself a surrounding envelope of aqueous vapor, larger than our own in proportion to his mass, and larger than that of all the planets together, the combined mass of which he exceeds by seven hundred and fifty times. We should also expect similar aggregations of aqueous vapors to surround all the fixed stars in proportion to their various masses, yet we do not find aqueous vapor there, but hydrogen instead. And in the distant telescopic nebulæ we still find hydrogen and nitrogen; even in the comets we find free hydrogen in vast predominance, but not free oxygen; so that we may roughly divide the bodies of stellar space into two grand categories,—those with atmospheres of hydrogen and those with atmospheres of oxygen.

It is true that the latter are limited to the planets of our own system, so far as direct observation goes, for we cannot see such dark planets as exist beyond our own solar system; but if such planets exist, as they must, for reasons stated later on, and revolve around their own central suns, we may infer, with the strength of demonstration almost, that if their suns correspond to our sun in this respect, their planets will correspond to our planets in a similar respect. But the bodies with atmospheres of oxygen are those which rotate around the sun substantially as a center, while with reference to themselves the sun is more or less a fixed body in space. It is true that our whole system is drifting through space, at present in the direction of the constellation Lyra, and directly away from that portion of space occupied by Sirius and Canopus, with an annual motion of probably hundreds of millions of miles. Professor Ball ("In the High Heavens") says, "In conclusion, it would seem that the sun and the whole solar system are bound on a voyage to that part of the sky which is marked by the star Delta Lyræ. It also appears that the speed with which this motion is urged is such as to bring us every day about 700,000 miles nearer to this part of the sky. In one year the solar system accomplishes a journey of no less than 250,000,000 miles." A speed of eight miles per second gives an annual rate of 252,288,000 miles. This speed, however, is greatly exceeded by many stars (as determined by displacement of the lines of the spectrum); the star No. 1830, of Groombridge's catalogue (see "In the

High Heavens"), has a rate of two hundred miles per second. The author says, "Indeed, in some cases stellar velocities are attained which appear to be even greater than that just mentioned. We do not, therefore, make any extravagant supposition in adopting a speed of twenty miles per second," which he takes as the average. "I have adopted this particular velocity as fairly typical of sidereal motions generally. It is rather larger than the speed with which the earth moves in its orbit." The distances, of course, are equally enormous. This author says, "The nearest star, as far as we yet know, in the northern hemisphere is 61 Cygni. . . . I think we cannot be far wrong in adopting a value of fifty millions of millions of miles. . . . In the course of a million years a star with the average speed of twenty miles a second would move over a distance which was about a dozen times as great as the distance between 61 Cygni and the solar system." This assuming that the solar system is at rest, which is not the case, as the author says, "Unless binary, stars do not remain in proximity, so far as we know; the general rule appears to be that of universal movement through space." This drift through space, however, no more affects the terms of the problem than the rotation of the earth upon its axis or its orbital motion affects the operations of an electric machine as the handle may be rotated to or from the direction of these motions. Both machine and reservoir of energy occupying a fixed relation with reference to each other, the positions of each are the same as though absolutely fixed. This is true of gravitation, likewise, as well as of all other natural and universal forces.

The fact established, then, that attenuated aqueous vapor is diffused throughout the interplanetary space occupied by our own solar system, and that it tends to surround our sun and planetary bodies with aqueous envelopes of increased density, proportionate to the action of gravity, the question arises, Is there any known force which will act through such interplanetary space to decompose such aqueous vapor into its constituent elements and deposit hydrogen gas around the sun and oxygen gas around the planets, and which, while maintaining a planetary temperature such as we find on the planets, will at the same time raise the hydrogen envelope of the sun to such a temperature of incandescence that it will become a glowing sphere of heated hydrogen, in which other constituents of the sun's mass will be raised to incandescence and partially volatilized in the intense heat of that incandescent gas; in which, in fact, the phenomena of the sun will become manifest? If so, two vastly important corollaries are inevitable: first, that the fixed stars, which also shine with the light of their own glowing hydrogen, are themselves surrounded by a similar aqueous vapor, diffused through their own adjacent space, and that, in consequence, not only our own planetary distances, but all interstellar space, as far as the utmost distance of the faintest fixed stars, is likewise pervaded by the same attenuated aqueous vapor, and that this is the grand source from which is derived all solar energy, not only of our own sun, but of all the other flaming orbs of space; and, second, which is still more important to us as citizens of the universe, that each flaming hydrogen sun must have surrounding it a correlative dark planetary system of its own, and that the complement of glowing hydrogen, as an incandescent envelope of the central orb, necessitates the corresponding supplement of cool oxygen as an envelope for each of such planetary bodies; in other words, that without such planets as our system possesses, there can be no suns such as our own and the other suns we see. Vast orbs might be conceived of as rotating in eternal darkness without associated satellites, but the incandescent atmosphere of hydrogen must have—not may have, but must have—subordinate planets substantially similar to ours, surrounded by atmospheres substantially similar to our own (for we find free nitrogen in comets, in meteorites, and in the faintest nebulæ), and these planets are thus fitted, so far as we can know, for the support of organic life and for the same orderly courses of nature as we see manifest around us. They must be cool, for at the planetary poles there must be a moderate temperature in contrast with the solar pole, which becomes, of necessity, highly heated; they must have an atmosphere of oxygen in order that the solar center may have an atmosphere of hydrogen; these planetary atmospheres must be supplied with nitrogen, because nitrogen is universally available, and similar causes operating under similar circumstances will produce like effects; these atmospheres must be charged with condensed aqueous vapors, and, if cool enough, must have deposited water in liquid form, for aqueous vapors when condensed by gravity are the correlated sources of supply of their respective gaseous components at both solar and planetary poles; and these planets must rotate in orderly periods around their central suns, or the aqueous vapors cannot be regularly and continuously disassociated into their elemental gases. These planets may be few or many—perhaps even a single one sometimes—for each sun, but they must be large enough or numerous enough to operate by their aggregate mass, so as to disassociate around the planets as much oxygen as their central sun disassociates of hydrogen in their combining proportions,—that is, two volumes of hydrogen for each one of oxygen. We will therefore find in such planets all the potentialities of life—we can see and study these planets, though physically invisible, as easily and as thoroughly as we do our own, for having the relationship of constitution between our own planets and our sun, we may thereby learn the essential relationship between any fixed star and its planets by directly studying the constitution of such star alone. Among the planets of our own system Neptune and Mercury, and those which exist adjacent to their boundaries, can be studied with difficulty and uncertainty; but what astronomer doubts that they are constituted much like the other planets, and have passed, or will pass, through such stages of progress as we find apparent among those more directly under our observation? While we shall thus find universality and harmony among all the starry systems, we shall not find identity; but with the guiding light of demonstrated scientific principles, we may apply our knowledge as a key to unlock the mysteries of the most distant stars. The Milky Way will gleam with new meaning, Sirius, Aldebaran, the Pleiades, will send us messages of fellowship, and the established sphere of creative energy will have expanded, with all its wondrous mechanism, to fill the universe. When we see at night a vast factory building with every window lighted, one who understands the operation and mechanism essential to the work of a mill sees not alone the illuminated windows, but the looms in motion, the flying shuttles, the spindles humming, the wheels turning, and all the complicated machinery in active operation. And he can even picture operatives at work in their various avocations, and the flashing windows, though themselves silent, are the visible index of the light within which illuminates and makes possible the work there performed. And so, when thus comprehended, the flaming stars, but points of light in the archways of the sky, themselves will reveal to us the wondrous workings within the realm which they illuminate and warm and vivify. We may also reasonably infer, as will be more fully explained further on, that there can be no actual basis for the opinion sometimes expressed, that great, dark, solid orbs —independent worlds, in fact—are drifting about through space at random, as it were, like homeless vagabonds. In these sparsely-occupied domains the head of each household, as in every well-regulated family, has all its different members gathered around in strict subordination, to aid in the support of the establishment. No sun no planets; no planets no sun, is the general statement of the sidereal formula. Like a sexual duality, the mutually correlated parts constitute a single, composite, and interdependent whole: one generates, concentrates, and transmits; the other receives, transforms, and delivers.

NOTE.—Regarding the absence of oxygen from the sun's atmosphere we quote the following from Lord Salisbury's very recent address (see note at end of Chapter I.): "It is a great aggravation of the mystery which surrounds the question of the elements, that, among the lines which are absent from the spectrum of the sun, those of nitrogen and oxygen stand first. Oxygen constitutes the largest portion of the solid and liquid substances of our planet, so far as we know it; and nitrogen is very far the predominant constituent of our atmosphere. If the earth is a detached bit whirled off the mass of the sun, as cosmogonists love to tell us, how comes it that in leaving the sun we cleaned him out so completely of his nitrogen and oxygen that not a trace of these gases remains behind to be discovered even by the sensitive vision of the spectroscope?" We shall find that the absence of oxygen in the solar envelope is a necessary corollary of its presence in those of the planets. The same is true, possibly, of nitrogen. Ammoniacal vapors are decomposable into hydrogen and nitrogen, and hydrocarbon gases into hydrogen and carbon, just as aqueous vapors are resolvable into hydrogen and oxygen. In the earlier stages of the earth's development we have abundant evidence of an atmosphere heavily laden with carbonic vapors, which have disappeared, to remain stored as fixed carbon, and the oxygen has also largely disappeared, to constitute the enormous mass of oxides in the earth's mass, while the nitrogen remains to dilute the remaining oxygen and constitute the air we breathe. Their common correlative, hydrogen, intermingled with metallic vapors, composes the vast atmosphere of the sun.

CHAPTER III.

THE MODE OF SOLAR ENERGY.

But is there such an available force? There is one, and only one,—electricity, when properly generated and suitably applied. It is an axiom of electrical science that any fluid which will at all conduct a current of electricity can be decomposed by a current of electricity. (See Urbanitsky's work, "Electricity in the Service of Man," Cassell's edition, page 154.) It is there stated (page 152), "We have frequently had occasion to mention certain chemical effects of electricity,—namely, the decomposition of gaseous compounds into simple gases." Page 157, "Whatever the substances we expose to the action of the galvanic current, decomposition takes place proportional to the strength of the current." Page 152, "Hydrogen is always evolved at the negative pole of the battery and oxygen at the positive pole. The gases can then be collected in different tubes, the hydrogen tube receiving twice as much gas as the oxygen tube; since water consists of two volumes of hydrogen and one volume of oxygen, it follows that the galvanic current decomposes water into its constituents. As chemically pure water has so great a resistance as almost to force us to consider it a nonconductor, it is generally acidulated with sulphuric acid. The smallest amount of acid diminishes the resistance considerably. The silent discharge is far more effective in bringing about this transformation than the spark discharge." Page 37, "Gases are bad conductors of electricity; if it had been otherwise, we should never have become acquainted with electricity, as it would have been conducted away by the air as fast as it was generated. The vacuum also does not conduct electricity, but *moist air* becomes a partial conductor. Moist air also will spoil the insulation of non-conducting supports. All bodies are more or less hygroscopic, and the moisture condensed on their surfaces *thus turns the best insulators into conductors.* Change of temperature also influences conductivity." Page 63, "When using induction machines, the moisture of the air often causes experiments to fail, especially before large audiences. The atmosphere becomes saturated with moisture, and it is often impossible to get the machine in working order." Several desiccating devices are mentioned by the authors of this work, as used with such machines, to prevent such dissipation or conduction of electricity from the machine into space by the aqueous vapor of the atmosphere. In describing the aurora borealis (page 93), these authors say, "The rarefied air is nearer the earth at the poles than the equator, in consequence of the earth's centrifugal motion, and, the earth being negatively electrified, negative electricity will flow from this point, directed against the *positively electrified upper layers of rarefied air.*" Same work, pages 127, 128, "The resistance (in liquids) diminishes as the temperature increases, a result which is exactly opposite to what occurs with metals. Conductivity for carbon increases with the temperature, thus agreeing with the action of liquids." Page 133, "To determine the resistance in liquids, the above methods cannot be employed, liquids being decomposed by the electrical current." Referring to the voltaic arc and the spark of the induction apparatus (page 200), it is said, "Dry air under great pressure offers a high resistance, but a *perfect vacuum is a perfect insulator*, and between these extremes there are degrees of rarification which admit of a flow of electricity." In general, it is said that electrical decomposition requires that the electrolyte be in liquid form, but this is not universally true, and throughout interplanetary space may not be true at all. In Ferguson's work on Electricity, it is stated that, "The passage of electricity through compound gases in a state of great rarity, as in the so-called vacuum tubes, frequently separates them up into their constituents." So, also, the opinion that electricity cannot be readily conducted through dry gases is refuted by the play of the auroral streamers. The distance from the surface of the earth of these electrical waves and the auroral arch is variously estimated at from seventy to two hundred and sixty-five miles, and in one instance "at a height of from four thousand to six thousand miles;" see article in Appleton's Cyclopædia. Certainly there could be no sensible moisture at the temperatures there prevalent, and especially at night and during the fall and winter months when these displays are very frequent. Whether the currents be due to induction, as between neighboring bodies one of which is electrified, or from direct emission, as in brush discharges, there must obviously be some medium of contact and continuity for the free transference of electrical energy through space. Regarding the *rationale* of electrolysis ("Electricity in the Service of Man"), after discussing certain other theories, the authors say, " Clausius, too, assumes an electrified condition of the molecules of each electrode, but he neither attributes to the galvanic current the force of direction nor power of decomposing. He points out that both the molecules of fluids and also their atoms are in continual motion. The atoms in molecules of fluids are held together but by a moderate force, and the molecules themselves constantly undergo changes both of synthesis and analysis. The galvanic current merely effects a regulated motion of the atoms; the positive ions are attracted by the negative electrode, and the negative ions by the positive electrode, and by this means are separated out from the liquid." Page 91, "The upper layers of air are more or less electrified, so as to have a potential differing from that of the earth, but *how their electrical condition has been produced is not at present known*. Condensation of water-vapor is supposed to produce electricity. Close to the earth the air has little or no electricity; the farther from the earth the greater the amount of electricity in the air." Referring to the sparking discharge, it is said, page 75, "The density of the air, however, has to be taken into account; the sparking distance is lessened in denser air, and becomes greater when the atmospheric pressure is diminished. Not only the density, but also the chemical composition of the medium influences the sparking distance. Faraday found the distances considerably less in chlorine gas, but *twice as long in hydrogen gas as in air.*" Page 74, "The sparking distance increases at a somewhat greater rate. than the difference of potential of the discharging bodies. . . . When the sparking distance becomes very great . . . it is proportional to the difference of potential." Page 91, "There is a difference of potential between the earth and points in the air above. In fine weather the potential is higher the higher we go, increasing usually at the rate of *twenty to forty volts for each foot.*"

It will be seen that, continued upward at this rate, the increased electrical pressure for each mile of elevation would be between 100,000 and 200,000 volts, or for each one hundred miles more than 10,000,000 volts; and at an altitude of one thousand miles, if carried so far, the potential would be between one and two hundred million volts, an electrical pressure quite inconceivable to us. Such a potential in currents of enormous quantity continually flowing from the earth to the sun would certainly decompose any aqueous vapors condensed around these bodies. But the question at once arises, What reason is there to suppose that such currents could possibly flow between the earth and the sun, across that vast intervening region of space, a distance of more than 90,000,000 miles? And would not the resistance to such currents in transit be so enormous that the entire potential, however great, would have been practically lost long before reaching the sun? To this there is a complete and irrefutable answer, not based upon any abstract theory, but upon established fact. It is an absolute certainty that electrical currents of enormous quantity and high potential are constantly passing between the earth and the sun, and that these currents have so free a passage—far more free than through any metallic circuits that we know of—that they pass over this enormous distance absolutely without appreciable resistance. We may note in this connection the well-known facts, now being largely utilized, though the art is still in its infancy, of telegraphing and transmitting all sorts of electrical currents over large distances without wires or any conductors, except those furnished by nature.

Of the currents between the earth and the sun, Professor Proctor, in his " Light Science for Leisure Hours," says, " Remembering the influence which the sun has been found to exercise upon the magnetic needle, the question will naturally arise, Has the sun anything to do with magnetic storms? We have clear evidence that he has. On the 1st of September, 1859, Messrs. Carrington and Hodgson were observing the sun, one at Oxford and the other in London. Their scrutiny was directed to certain large spots which at that time marked the sun's face. Suddenly a bright light was seen by each observer to break out on the sun's surface and to travel, slowly in appearance, but in reality at the rate of about seven thousand miles in a minute, across a part of the solar disk. Now, it was found afterwards that the self-registering magnetic instruments at Kew had made *at that very instant* a strongly-marked jerk. It was learned that at that moment a magnetic storm prevailed in the West Indies, in South America, and in Australia. The signal men in the telegraph stations at Washington and Philadelphia received strong electric shocks; the pen of Bain's telegraph was followed by a flame of fire; and in Norway the telegraphic machinery was set on fire. At night great auroras were seen in both hemispheres. It is impossible not to connect these startling magnetic indications with the remarkable appearance observed upon the sun's disk. But there is other evidence. Magnetic storms prevail more commonly in some years than in others. In those years in which they occur most frequently it is found that the ordinary oscillations of the magnetic needle are more extensive than usual. Now, when these peculiarities had been noticed for many years, it was found that there was an alternate and systematic increase and diminution in intensity of magnetic action, and that the period of the variation was about eleven years. But at the same time a diligent observer had been recording the appearance of the sun's face from day to day and from year to year. He had found that the solar spots are in some years more freely displayed than in others, and he had determined the period in which the spots had successively presented with maximum frequency to be about eleven years. On a comparison of the two sets of observations it was found (and has now been placed beyond a doubt by many years of continual observation) that magnetic perturbations are most energetic when the sun is most spotted, and *vice versa*. For so remarkable a phenomenon as this none but a cosmical cause can suffice. We can neither say that the spots cause the magnetic storms nor that the magnetic storms cause the spots. We must seek for a cause producing at once both sets of phenomena." It will be observed that the phenomena seen in the sun were marked *at the same instant* by violent electric perturbations on earth. Hence something must have passed with the velocity of light, which we know to be at the rate of 188,000 miles per second, or in about eight minutes from the sun to the earth. But it is stated in "Electricity in the Service of Man," page 82, that, "According to the theoretical calculations of Kirchhoff, as well as of Ayrton and Perry, the velocity of electricity in a wire *without resistance would be equal to the velocity of light.*" Hence we perceive that the apparent difficulty has vanished in the light of observed fact, and that currents of electricity do pass and are constantly passing between the earth and the sun without the slightest loss of speed,—that is to say, without resistance. We shall find in the sequel that the above phenomena were caused most probably by a partial interruption of a constant direct current from the earth to the sun, instead of by an opposite return current from the sun to the earth. In further illustration of the above facts we quote the following, page 172, " Electricity in the Service of Man:" " Many attempts have been made to find a connection between the spots and prominences in the sun and the electrical phenomena on the earth. Professor Forster says that by numerous magnetic observations of the last thirty or forty years it has been proved that the formation of black spots on the surface of the sun, and the generation of pillars and clouds of glowing gases in the immediate neighborhood of the sun, stand in close connection with certain deviations in direction and intensity of the earth's magnetic forces." Professor Proctor, in his " Light Science for Leisure Hours," says, "From all this it appears, incontestably, that there is an intimate connection between the causes of auroras and those of terrestrial magnetism. . . . The magnetic needle not only swayed responsively to auroras observable in the immediate neighborhood, but to auroras in progress hundreds and thousands of miles away. Nay, as inquiry progressed, it was discovered that the needles in our northern observatories are swayed by influences associated even with the occurrence of auroras around the southern polar regions. . . . Could we only associate auroras with terrestrial magnetism, we should still have done much to enhance the interest which the beautiful phenomenon is calculated to excite. But when once this association has been established, others of even greater interest are brought into recognition; for terrestrial magnetism has been clearly shown to be influenced directly by the action of the sun. . . . We already begin to see, then, that auroras are associated in some mysterious way with the action of the solar rays. The phenomenon which had been looked on for so many ages as a mere spectacle, caused perhaps by some process in the upper regions of the air of a simple local character, has been brought into the range of planetary phenomena. As surely as the brilliant planets which deck the nocturnal skies are illuminated by the same orb which gives us our days and seasons, so are they subject to the same mysterious influence which causes the northern banners to wave respondently over the starlit depths of heaven. Nay, it is even probable that every flicker and coruscation of our auroral displays correspond with similar manifestations upon every planet which travels round the sun." In Professor Ball's late work, "In the High Heavens," the author says, "Dr. Schuster suggests that there may be an electric connection between the sun and the planets. In fact, with some limitations, we might even assert that there *must* be such a connection. It is well known that great outbreaks on the sun have been immediately followed, I might almost say accompanied, by remarkable magnetic disturbances on the earth. The instances that are recorded of this connection are altogether too remarkable to be set aside as mere coincidences. Dr. Huggins has not referred in this connection to Hertz's astonishing discoveries; but it seems quite possible that research along this line may throw light on the subject, *at present so obscure,* of the electric relation between the sun and the earth." Of this common electrical relationship between our sun and the different planets, and of these with each other, Professor Proctor says, in his article, "Terrestrial Magnetism," "Interesting as are the bonds of union which Copernicus and Kepler and Newton have traced in the relations of our system, *it would seem as though we were approaching the traces of a yet more wonderful law of association.* We see the earth's magnetism responding to the solar influences, not merely in those rhythmic motions which belong to the periodic variations, but in sudden thrills affecting the whole framework of our globe. The magnetic storms which are called into action by such solar disturbances as the one of September, 1859, are, we may feel sure, not peculiar to our own earth. The other planets feel the same influence,—not, perhaps, in exactly the same way, but according to the constitution and physical habitudes which respectively belong to them. So that one can scarce conceive a subject of study at once more promising and more interesting." Of these prophetic shadows which science often seems to cast before, Professor Nichol, in his "Architecture of the Heavens" (referring to Sir William Herschel), says, "Without difficulty or pretence he there casts aside an idea which had not been questioned before, unless in a few of those obscure, indefinite speculations *which, strangely enough, often prelude important discoveries.*" These facts are thus incontestably established: that electric currents of enormous energy and vast quantity

THE MODE OF SOLAR ENERGY. 81

are constantly passing without appreciable resistance and with the speed of light between the earth and the sun; that such currents cannot be conducted through vacua, or through dry gases, or through a dense medium; and that, whatever other matter may exist in the intervening space, such space is pervaded throughout by an attenuated vapor of such constitution and density that it will transmit such electrical currents with the highest conceivable efficiency. We know that such passage of these currents cannot depend upon the ether of space which is acted upon by the sun to produce the ethereal undulatory vibrations of light and heat, for, after we have produced the most perfect vacuum possible, we find that the rays of light continue to pass through it as freely as they pass through space, while currents of electricity cannot be made to pass at all. Hence we know to a certainty that the medium which transmits these enormous currents of electricity must be a vapor capable of conducting electricity, that it must hence be decomposable by the electric current, and that when decomposed one of its elements must consist of hydrogen gas and the other of oxygen; in other words, that this conducting medium must consist of attenuated *aqueous* vapor, commingled doubtless with other vapors which themselves, like the acid of the acidulated water used in electrolysis, aid in the conduction of these enormous currents. We also know that such vapors in space will be necessarily attracted, by gravitation, around the solar and planetary bodies immersed therein, and must form

f

condensed vaporous atmospheres or cloud masses, and if these are decomposed by the passage of such currents of electricity, that hydrogen gas will be liberated at the solar galvanic pole and oxygen at the terrestrial or other planetary pole, precisely as we find to be the case in nature. Will such gaseous envelopes, then, have the same temperature for each gas when thus liberated, or will the hydrogen envelope of the sun be heated to incandescence, due to the passage of the electrical current?

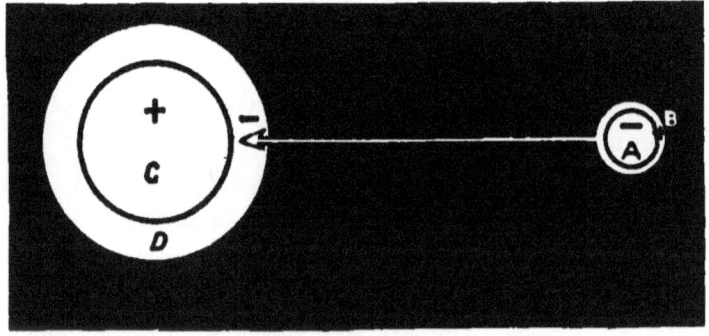

Electrical polarities of sun and planets. A, body of the planet; B, planetary electrosphere; C, body of the sun; D, solar electrosphere.

The temperature of interplanetary space is probably very low. Of this Professor Ball says, "What this may be is a matter of some uncertainty, but from all the evidence available it seems plain that we may put it at not less than three hundred degrees below zero;" and the same author adds, "The temperature is taken to be sixty-four degrees below zero, being presumably that at the confines of the atmosphere." Whatever the temperature of space, or its variations, may be, the passage of the planetary electricity through the condensed hydrogen envelope of the sun will produce great changes in the heat of that body and of the solar core within. While with a small electrolytic apparatus we find no special differences of temperature in the gases, with large quantities of electricity, driven at a high potential, we find that a new and startling result ensues. Something of this sort is seen in the operation of electric arc-light lamps, now in common use, in which two slightly separated carbon points are traversed by a current of considerable potential. The current is driven across the intervening space between the points, carrying with it an atmosphere of disintegrated carbon, through which the electricity is carried at its highest speed, and a most brilliant light is produced. In "Electricity in the Service of Man," page 151, it is said, "We may conclude from this that the current does not cease when the arc of light is formed. The resistance of the arc seems to be only very slight; in fact, the current must be conducted by it." Of the structure and constitution of the luminous electrosphere, or arc, produced in these lamps, " Professor J. A. Fleming," says the *Scientific American,* " has shown that the well-known color of the light of the electric arc from carbon points is due to the incandescence of the carbon filling the space between the positive and the negative rods. The true arc is here, and exists in a space filled with the *vapor of carbon*, which has a brilliant violet color. Examined by the spectroscope, the central axis of the carbon arc gives a spectrum marked by two bright violet bands. Outside this is an aureole of carbon vapor of yellow or golden color. The electrical strain of the arc occurs chiefly *at the surface of the crater* which forms at the end of the positive rod, where, in fact, the principal work of generating light is done; for *eighty per cent. of the total light of the arc comes from the incandescent carbon at this place.* Thus, in a sense, the arc light is mainly an incandescent light, the effect being produced by the layer of carbon which is being constantly evaporated at an extremely elevated temperature. Hence the light of the carbon arc is not, and can never be, white, as it is sometimes described as being, but must always be tinted violet by the carbon vapor normally present between the rods."

The significance of the above-quoted extract will be readily perceived when we come to consider the action of the direct planetary electrical currents upon the solar envelope, the effects in both cases being substantially identical. The quantity and intensity of the electric current, as it passes through the incandescent arc to the negative pole, and thence back to the dynamo, are diminished exactly in proportion to the energy expended in the generation of the light and heat of the arc. It is precisely the same as in the operation of a turbine water-wheel; if working at its highest efficiency, the discharged water is almost deprived of force: its gravity has been converted into work. In the electric light this conversion is only partial, owing to atmospheric and other conditions; but in the case of the solar envelope and its core, it is nearly, if not altogether, perfect, so that the currents of electricity are almost entirely converted into light and heat, or expended in the electrolytic decomposition of the surrounding aqueous vapors, and do not reappear as electricity, but as converted solar energy. Brilliant, however, as the light rays are in a powerful arc lamp,—perhaps the nearest to solar light we can produce,—the obscure heat rays are far more numerous and powerful. On page 476 of the work just cited a table is given, showing the proportion of visible and invisible rays emitted by different illuminants, and with the electric lamp, even, ninety per cent. of all the rays emitted by the voltaic arc are heat rays, which are obscure and invisible. But the startling effects of electricity of large quantity and high potential, in the decomposition of water, are far more strikingly exhibited by an apparatus shown in 1893 at the Chicago Exhibition by a firm from Brussels, and which is described in the *Electrical Review* as follows: "An ordinary wooden pail is three-quarters filled with water slightly acidulated; a lead plate about nine inches broad by sixteen inches long dips to the bottom of the pail and is connected to an incandescent dynamo machine capable of giving over one hundred and fifty ampères. The iron rod, or article to be heated, is connected to the pole of the dynamo and simply dipped into the water; it immediately becomes heated and rapidly rises to a melting temperature; only that portion of the metal completely immersed becomes heated, and the heating is so rapid that neither the water nor that portion of the metal out of the water becomes very warm. Wrought iron and steel actually melt if long enough held under water. A carbon rod subjected to this process becomes amorphous carbon, proving that a temperature of at least four thousand degrees Centigrade has been reached, and it is stated that with two hundred and twenty volts' pressure a temperature of eight thousand degrees Centigrade has been reached. There are various theories to account for this phenomenon, but from close observation it appears to be a case of arc heating. The moment the metal is plunged into the water *it is enveloped in hydrogen gas* decomposed from the water. This envelope of gas parts the water and metal, forming an arc, which raises the surrounding gaseous envelope to an enormous temperature; the metal surrounded by this arc is almost immediately raised to the same temperature. *A flame of burning hydrogen* appears around the metal on the surface of the water. The principle of the method is the same as that on which the burning of an arc light between two carbon points under water depends. An arc lamp will burn quite steadily under water if the connections are made water-proof; the arc itself requires no protection."

It will be seen that the process above described is precisely analogous to that involved in the problem of the sun's energy. The planets correspond with the leaden plates, upon which oxygen is disengaged from the water, while at the same moment the liberated hydrogen necessarily appears at the opposite pole. The generation of hydrogen gas forms an envelope or atmosphere of hydrogen around the sun which forces back the aqueous vapor. The current, in passing through this gaseous envelope to the metal core within, intensely heats the hydrogen, which rapidly communicates its rising heat to the central core. If this core is composed of metals, and the temperature be raised sufficiently high, which only depends upon the quantity and working pressure of the electricity employed, the metal core will be volatilized in whole or in part, and, if of mixed metals, we will find the presence of these elements revealed in the spectroscopic lines corresponding thereto, and the flames and flashes of hydrogen at the surfaces beyond the envelope, at the surface of contact with the matter of space, will be also seen. In fact, such an experiment, properly prepared, could be made to show roughly most of the phenomena of solar light and heat as they actually appear, such as sun-spots, prominences, jets, plumes, faculæ, the photosphere, chromosphere, absorption bands, vortical disturbances, metallic vapors, and the complete solar spectrum, with the different Fraunhofer lines. In the case of the sun, these currents must be measured by millions of ampères, and possibly by hundreds of millions of volts, instead of by mere hundreds, while the hydrogen envelope extends outward from the sun's surface hundreds of thousands of miles until, perhaps, finally merged into the corona. As the currents pass from the planets and planetoids (for not only the larger planets, but all the planetary bodies of our system must contribute, if any of them contribute) to the sun, or rather to the sphere of its electrical action, without resistance, so long as these planets generate constant currents of the same, or nearly the same, potential, so long will the sun maintain his constant light and heat; if these are increased or diminished, the sun's light and heat will be temporarily, but only temporarily, increased or diminished; and this process must continue, without further loss or change, indefinitely into the future. Whatever the sun may gain by increment of meteoric masses may pass for what it is worth, but the gradual contraction of his volume cannot proceed while his present temperature is maintained by the passage of such currents, —that is to say, his light and heat will remain constant, and also his mass and volume, so long as the electric currents which pass from the planets to the sun and the constitution of space which surrounds the sun and planets themselves remain constant.

It now remains to consider how such enormous currents of electricity can be generated and maintained. We know, of course, that chemical changes cannot operate to produce them. They must be derived from something contained in or diffused through interplanetary space, and the planets themselves must be the means by which such currents of electricity are brought into effective operation. On our own earth we have many kinds of mechanically-constructed electrical apparatus which *generate* electricity, to use a popular expression, or which, more properly, separate the opposite potentials from an unstable electrical tension or equilibrium of the matter of space. These machines practically take positive electricity from the mutually-balanced electric potentials of which the earth and its surrounding gaseous envelope are the vast common storehouse, in such manner that the positive electricity thus drawn out from and again passing into the common storehouse shall, during such transit, be compelled to pass through channels which will cause it to do work, at the expense of its

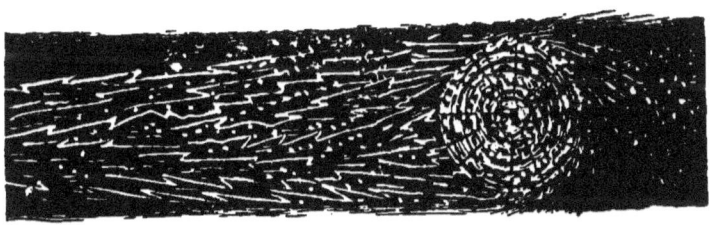

Ideal view of the generation and transmission of planetary electricity.

potential or pressure, during its passage, or in which electricity is raised in its electro-motive force from a lower to a higher potential or pressure, just as the pressure of water is increased when delivered from a greater or a still greater height, or steam, when confined in space under higher and still higher temperatures. But none of these machines actually *generate* electricity *ab initio;* they merely put into effective operation the pre-existing force. The mass of the earth is of irregularly negative polarity, the air above is positive, and as we ascend, the potential, or voltage, or pressure increases at a nearly uniform rate of from twenty to forty volts for each foot. The earth is thus surrounded by an electrosphere as well as an atmosphere, and the two are not coincident, for while the pressure of the atmosphere diminishes as we ascend, that of the electrosphere increases. The moon, too, and each planet must have its electrosphere, and around the sun's core we can see the solar electrosphere in its visible glory. Thus, all our planets rotate upon their axes and revolve around the sun, each surrounded by an enormous electrosphere, just as an electrical induction machine is surrounded, when in operation, with an electrosphere of its own, and which, by breaking connection with the conductor which carries away its current, becomes, when shown in a darkened room, clearly visible. In "Electricity in the Service of Man" it is said, page 63, "The inductive action of the machine is quite as rapid and as powerful when both collectors are removed and nothing is left but the two rotating disks and their respective contact or neutralizing brushes. The whole apparatus then bristles with electricity, and if viewed in the dark presents a most beautiful appearance, being literally bathed with luminous brush discharges." This is a true aurora.

Let us now examine some of these more recent electric machines,—the later induction, not the older frictional machines, for it is obvious that the rotation of the planets, if they operate as electric generators, or separators, must act by induction

THE MODE OF SOLAR ENERGY. 91

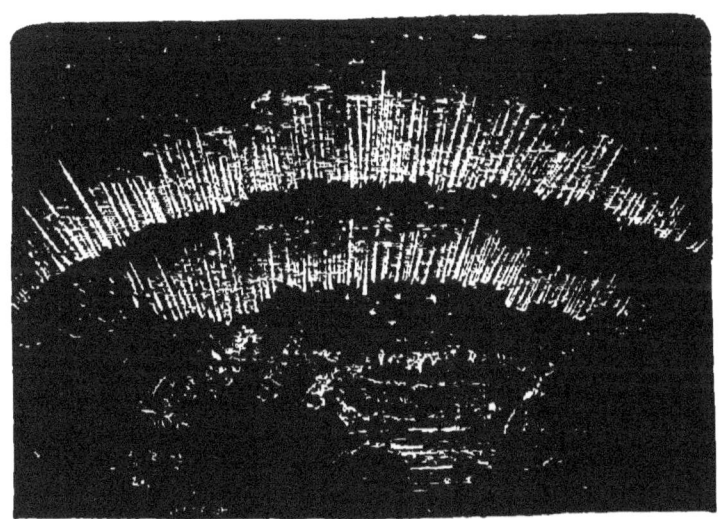

The Aurora Borealis. (From "Electricity in the Service of Man.")

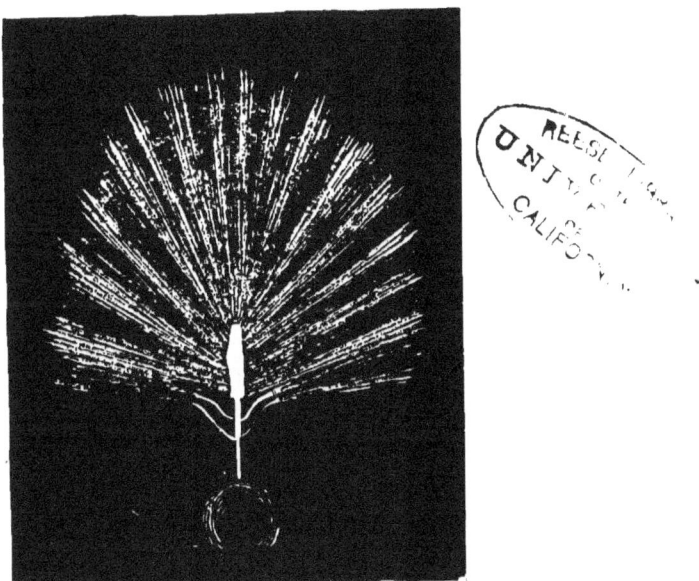

Diffused brush discharge of electrical machine, when operating with its current cut off or interrupted between machine and principal condenser.

and not by friction. The frictional machines are of the old type and are well known from the books; in these a glass disk or cylinder is rubbed upon in its rotation by an amalgamated (so called) friction pad fixed securely to the bed of the machine. But more recently these have been replaced by far more powerful and simple machines which operate entirely by induction, like approaching thunder-clouds, for instance, and in which one or more glass disks are merely rotated rapidly and freely in the air, these disks having a number of light metallic sectors, such as bits of tin-foil, pasted on their outer sides at equal radial intervals, and with metallic collecting brushes which, however, barely graze the surfaces of the rotating disk. There is no pressure and no friction, except that of the disks as they freely revolve in the atmosphere.

In the above-quoted work, page 61, is a description of Wimshurst's influence machine, one of the most recent and most powerful, which we condense as follows: This machine was produced about 1883. It consists of two circular disks of thin glass fourteen and one-half inches in diameter in the sample described, attached at their centers to loose bosses, so as to be rotated by cords and pulleys operated by a handle, in opposite directions. The disks rotate parallel with each other and are not more than one-eighth of an inch apart, and have their surfaces well varnished; and attached by cement to their outer surfaces are twelve or more radial, sector-shaped plates of thin brass- or tin-foil, disposed around the disks at equal distances apart. These sectors take the place of the "inductors" of Holtz's instrument, and appear to act also as carriers, though the exact nature of their action is somewhat mysterious. It appears, however, probable that those acting for the time as carriers on the one disk act at the same time as inductors on the other. The two sectors on the same diameter of each disk, at opposite sides of the center, are twice in each revolution momentarily placed in metallic connection with one another by means of a pair of fine wire brushes attached to the ends of a bent metal rod loosely pivoted at the center of each disk, the metal sectors *just grazing* the tips of the wire brushes as they pass. There is one of these bent rods on the outside of each disk, and their position as pivoted on their center can be varied at will, both with reference to the one on the opposite side and to the position of the fixed collecting combs. The efficiency of the machine varies with their position, and the maximum appears to be generally when the brushes touch the disks on diameters crossing the position of the collecting combs at about forty-five degrees, and with the bent rods on opposite sides at right angles to each other. The collecting combs are simple forks with collecting points turned inward, which forks embrace the opposite sides of the disks outside, which freely rotate between them, and they are supported on insulated posts. These supports may be small Leyden jars or condensers, with discharging knobs, or may be connected with similar condensers at a distance, or arranged in batteries or otherwise. The presence of the collecting combs is not necessary to the operation of the machine, their sole function being to carry away the positive electricity as generated. The machine is self-exciting, and it is believed that the *initial action* must be due to friction in the layer of air contained between the plates, which, as above stated, are only about one-eighth of an inch apart. It is nearly independent of atmospheric conditions, and not liable to reverse its polarity, as are the Voss machines. The Voss machine uses a larger glass disk which does not rotate, but is fixed, and which has a central opening three inches wide, with a different arrangement of tin-foil disks or sectors, and a smaller glass disk rotates parallel with it. The Holtz machine is somewhat similar, using a single rotating, well-varnished glass disk revolving opposite a well-varnished larger disk, the latter provided with three sector-shaped openings or windows, with varnished paper inductors or flaps passing through these windows so as to touch the revolving disk. There are also two series of fine metal points held by brass bars provided with insulated handles and discharging knobs.

It is only necessary to give a general idea of the construction and operation of such machines, as their specific construction can be readily learned from the books. Of the mode of operation, however, it is said, "What takes place when the machine is in action is of a very complicated nature, and can hardly be said to be perfectly understood." With a Wimshurst machine having disks of a diameter of fourteen and one-half inches " there is produced under ordinary atmospheric conditions a powerful spark discharge between the knobs when they are separated by a distance of four and one-half inches, a pint size Leyden jar being in connection with each knob (one on each opposite diameter of the two disks), and these four-and-one-half-inch discharges take place in regular succession at every two and a half turns of the handle. It is usual to construct the machine with small Leyden jars or condensers attached to conductors, by which the spark is materially increased. A machine has been constructed with plates seven feet in diameter, which, it was believed, would give sparks thirty inches long; but no Leyden jars have been found to withstand its discharge, all being pierced by the enormous tension." Three of Toepler's induction machines (see page 59, "Electricity in the Service of Man"), connected together, gave a current which maintained a platinum wire one-fifth of a millimeter thick continually at a red heat, and was also capable of decomposing water.

CHAPTER IV.

THE SOURCE OF SOLAR ENERGY.

The remarkable resemblance between the mode of operation and effects of these electrical induction machines and the vast rotating electrosphere of the earth must be at once apparent. The operation is precisely the same, and the results must, *pari passu*, be substantially similar. We need not seek for precise parallelism of structure, because these machines themselves, it has been shown, widely differ in structure among themselves. But the almost infinitely more vast terrestrial electrosphere, which cannot be less than ten thousand miles in diameter, and perhaps much more (if we may form an opinion from the relative magnitude of the field of action of the hydrogen envelope which constitutes the solar electrosphere), rotating in the attenuated vapors of space, among which vapors that of water plays a most important part, and which vapors constantly impinge with various disturbances of contact against the more and more attenuated layers of the terrestrial atmosphere, and which gradually, from within outward, less and less partakes of the earth's rotation until, finally, its rotatory movement is lost in the vast ocean of space, establishes the certainty that enormous quantities of electricity must there be disengaged, precisely as in the machines which we have described, and to learn the potential or active pressure of this electricity we have only to consider the fact that we find a rise so rapid, as we ascend through our atmosphere, that the potential increases by from twenty to forty volts for each foot. That these currents are transmitted to the sun without appreciable resistance we already know, and that they are there transformed into light and heat we can, from the previously cited experiments, see.

But it may be urged that the resistance of such attenuated vapors in space, and the generation of electricity in such quantities, would inevitably retard and finally destroy planetary motion. The sufficient answer to this is found in the consideration that the same facts must exist under any possible mode of organization of our solar system, and that such interference, besides, must have absolutely prevented its formation at all, if such were the case. All the matter of our planetary system together is only one seven-hundred-and-fiftieth that of the sun; if this were added to the sun's bulk it would but slightly enlarge it. But all this solar and planetary matter together, if distributed over the space occupied by our planetary system,—and, by the nebular hypothesis of the organization of our solar system, this is requisite,—and having an axial diameter one-half that of its equatorial (see Proctor's "Familiar Essays on Scientific Subjects,"—"Oxygen in the Sun"), would have had a density of only about one four-hundred-thousandth that of hydrogen gas at atmospheric pressure. This nebular mass must have had a diameter at least sixty times that of the distance of the earth from the sun and a depth of thirty times its distance. That this enormous mass of attenuated matter should ever have been made to rotate as a whole by any force of attraction, repulsion, or rotation, with a tenuity so great that, if measured by an equal volume of hydrogen gas, —the lightest substance known to us,—it would have furnished material for four hundred thousand such systems as ours, presupposes a resistance so slight that the planets themselves, when coagulated out of such a mass, could never in any conceivable time exhibit retardation from such a source; and we know to a certainty that such attenuated vapors do exist in space, for electricity cannot be transmitted through a vacuum, and it is transmitted with perfect freedom between the earth and the sun. But it may be said that the laws were then different. If they were different then, they are doubtless different now. If, on the other hand, we assume that the bodies of which our solar system is composed were simply aggregated into concrete masses from meteoric dust, the difficulty is not lessened; for if the resistances to their operation now are such as to perceptibly retard their motions, they must have operated still more powerfully to originally prevent them; while, if hurled forth by an almighty fiat, complete from the hand of creative energy, the same force which impelled them forward must have also established the laws under which they now move.

It is calculated that our earth must be losing

time, by tidal retardation, at the rate of one-half the moon's diameter in each twelve hundred years (see Proctor, "Light Science for Leisure Hours,"—"Our Chief Timepiece Losing Time"), and that "the length of a day is now more by about one eighty-fourth part of a second than it was two thousand years ago." Perhaps, however, we may discover that these changes are themselves periodic and increase in cycles to a maximum, and then diminish, as is the case with magnetic, planetary, and stellar variations, and other similar changes, when sufficiently long observed; for while such changes may very well accompany a theory under which our system and all other systems are slowly running down to decay and death, it is entirely incompatible with the primal forces under which they *must* have been originally formed. In other words, if the tides are dragging back our earth without compensation, this dragging back can only come from the oceanic deposit of water on the earth from the aqueous vapors of space which do not partake of the planetary rotation and orbital movement of the earth. But if these can now retard the earth's motion, they must have originally prevented it in the beginning. This loss of time is, moreover, merely inferential from mathematical computations, and its basis is found in the belief that all the operations of nature are in a slow process of degradation, and the calculated loss itself may be merely theoretical, and not true in fact. Professor Proctor himself concedes the uncertainty of this alleged retardation when he says in the same article, "At this rate of change our day would merge into a lunar month in the course of thirty-six thousand millions of years. But after a while the change will take place more slowly, and *some trillion or so of years* will elapse before the full change is effected."

While the processes of nature are generally believed to be running down, everything is bent to that belief; but the forces of nature must, nevertheless, be uniform and supreme, for it is by these forces that the expected results are to be achieved. That changes occur constantly is inevitable, but the source of these must be looked for in the interaction of original forces, and not in the degradation of systems. There is reason to believe, in fact, that the repulsion of the terrestrial electrosphere by that of the moon may itself be sufficient to counteract such retarding force of lunar gravity, for the tides upon earth are not merely oceanic, but atmospheric, and on the latter the electrical repulsion of the moon must act very powerfully and with directly counteractive effect.

Let us now apply the preceding principles to the problem under review. All planetary space is pervaded with attenuated vapors or gases, among which aqueous vapor occupies a leading place. The planets and all planetary bodies, having opposite electrical polarity from the central and relatively fixed sun, by their orbital motions around and constant subjection thereto act as enormous induction machines, which generate electricity from the ocean of attenuated aqueous vapor, each planet being sur-

THE SOURCE OF SOLAR ENERGY. 101

rounded by an enormous electrosphere, carried with the planet in its axial and orbital movements, the successive atmospheric envelopes gradually dimin-

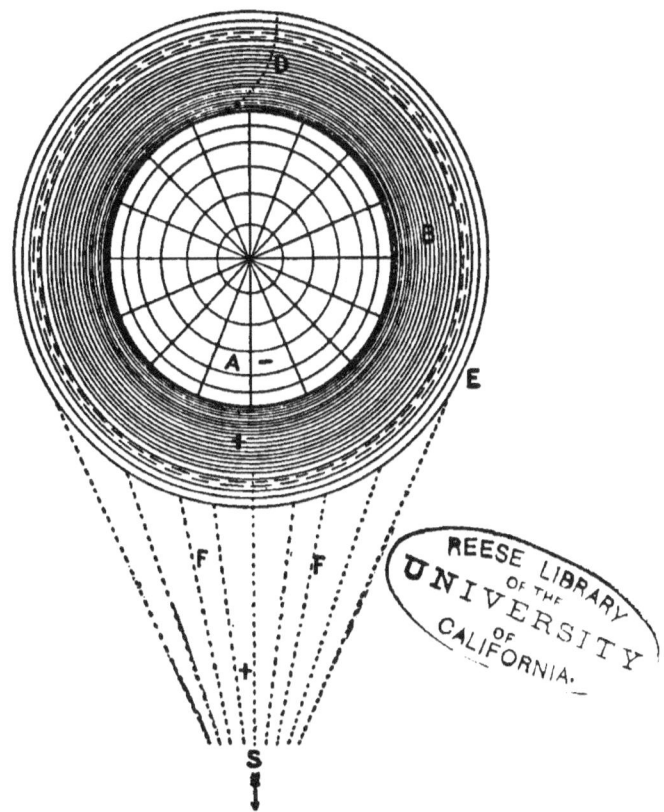

Planetary generation and transmission of electrical energy.—A, the planet; B, electrosphere showing circles of gradually diminishing rotation; E, interplanetary space; D, curve of gradually diminishing rotation; F, F, currents of electricity flowing to the sun; S, direction of the sun.

ishing in rotational velocity until merged into the outer ocean of space. As the planets advance in their orbits they plunge into new and fresh fields, and, as the whole solar system gradually moves

9*

onward through space, these fields are never reoccupied. These electrospheres, by their rotation, generate enormous quantities of electricity at an extremely high potential,—so high that we can scarcely even conceive it,—and this electricity flows in a constant current to the sun, where it disappears as electricity, to reappear in the form of solar light and heat. These planetary currents also flow towards such other negatively electrified bodies as may exist in space—the comets and fixed stars, for example—in proportion to their distance; for, since resistance is not appreciable between ourselves and the sun, as is also the case with light, so, like light, our electricity must pass outward as well as inward to take part in the harmonious operations of the whole universe. But it should be noted that the distribution of electric energy in the form of currents is quite different from that of light or other radiant energy; for while light is diffused from a center outward through space, electric currents, on the contrary, are concentrated and directed along lines of force to concrete centers of opposite polarity. As a consequence, the intensity of light decreases according to the squares of the distances traversed plus the resistance to the passage of the light itself, while the electric current is only diminished by the resistance of the medium through which it passes. As the light of the sun has a velocity of one hundred and eighty-eight thousand miles per second, and the electric current between the earth and the sun the same, it will be seen that the resistance is practically alike for these two forms of energy. Indeed, the striking resemblance between the ethereal vibrations which constitute light and heat and exceedingly rapid alternating currents of electricity through molecular media may suggest

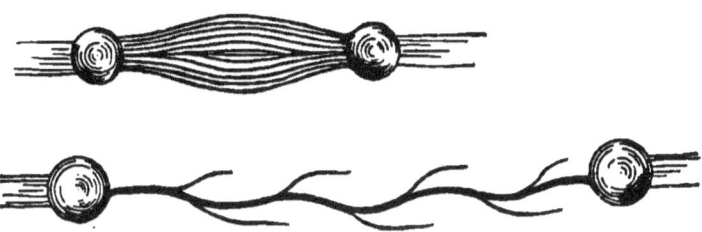

Upper figure.—Gradual discharge between two conductors, in partial vacuum.
Lower figure.—Sudden electric discharge through the atmosphere, from left to right.

that the transformation of one force into the other is some sort of a "step-up" or "step-down" process, much higher in degree, but of the same character as the well-known analogous electrical transformations used in the arts. It should also be borne in mind that, while the *intensity* of light diminishes according to the above law, the *quantity* remains the same, less resistance, as the area covered increases precisely in the same proportion as the intensity diminishes,—that is, in the ratio of squares.

Around the earth and other planets gravity attracts the aqueous vapors in increased density, the same as around the sun; but the electric currents passing between the planets and the sun decompose this aqueous vapor into its constituent gases, hydrogen and oxygen. The oxygen is deposited within the positive electrospheres of the planetary bodies, where it mingles with nitrogen to form our atmosphere and those of the other planets. In this float the aqueous vapors condensed from space, which are lighter than air. (See Tyndall, "The Forms of Water:" "It also sends up a quantity of aqueous vapor which, being far lighter than air, helps the latter to rise.") These aqueous vapors, condensed into clouds and precipitated upon the earth, form our oceans and their affluents. The hydrogen gas disengaged upon the sun's surface forms a similar envelope, which is penetrated by the planetary electric currents, and is thus highly heated and rendered incandescent; the glowing hydrogen transmits its heat to the sun's mass within, which is thus raised to, and permanently maintained in, a liquid or densely gaseous state, its metallic constituents being volatilized in part, and these metallic vapors mingle with the lower strata of hydrogen to form the sun's photosphere, while, above, the glowing hydrogen grows more pure, and finally, at a distance of hundreds of thousands of miles, is merged into the corona, which is composed, in part at least, of cosmical dust rotating around and repelled by the sun, and which shines partly by reflected light, partly by that of the relatively cooler hydrogen, and partly, perhaps, by electrification of its constituents by the powerful currents passing through it. Each of the planetary bodies, large or small, takes its proportionate part in the generation and transmission of electricity, according to its volume, mass, and motion. As an adjunct to this electrical sequence we have learned that any interruption of such currents between the generator and the receiver will cause the generating apparatus to glow with diffused electrical light, as is the case with the Wimshurst machine already described. When such connection is removed, it is said, "the whole apparatus bristles with electricity, and if viewed in the dark presents a most beautiful appearance, being literally bathed with luminous *brush* discharges." Such a phenomenon recalls at once the aurora borealis; and when we find this as a sequence of the electrical storm of the first of September, 1859, before described ("at night great auroras were seen in both hemispheres"), and connect with this the persistence of electricity upon insulated surfaces (see "Electricity in the Service of Man," page 53: "Glass being a bad conductor, the electricity does not spread all over the plate, but remains where it is produced"), we shall inevitably conclude that there was some partial interruption in the current flowing from the earth to the sun at that moment; and if we recall that at that very instant "suddenly a bright light was seen by each observer to break out on the sun's surface and to travel across a part of the solar disk," we shall learn that the processes connected with the production of such a bright light will interrupt in part the terrestrial current. We can readily understand that if this bright light exceeded in electrical intensity that due to the earth's current, it might temporarily reverse the polarity of the afferent current or retard its flow, like the so-called "backwater" of a mill. It would be like attempting to discharge steam at sixty pounds' pressure into a vessel filled with other steam at sixty-one pounds. Whence, then, came this bright light? Perhaps from the conjoint action of some other planet, perhaps from sudden chemical disassociation beneath the surface, perhaps by the abnormal piling up of depths of transparent glowing hydrogen or other local disturbance.

And this leads to the consideration of the uniformity of solar action. The planetary electrospheres will be constant in their operation if the constitution of surrounding space remains uniform; but we shall find reason to believe that there are currents in the ocean of space, as there are currents in our own seas, and electrical generation will necessarily vary when such currents are encountered. The sun itself in such case, however, will become an automatic regulator, for his density being but one-fourth that of the earth, and the spectroscope having shown his chemical composition to a large extent, we know that his mass must be either liquid or vaporous, and perhaps in part both. Such masses readily respond to variations of temperature, expanding as it rises and contracting as it falls. Hence, if a portion of space were reached where the action of the planetary electrospheres was increased by relative increase of temperature in some interstellar "Gulf Stream," the sun's volume would expand and compensation be at once established, while, conversely, with diminution of such planetary action, the solar volume would contract and an increased supply from his reserve store be given out thereby. In this way the condensation relied upon to give us heat for

THE SOURCE OF SOLAR ENERGY. 107

seven or seventeen million years becomes a compensating mechanism, self-operative through the most distant cycles of time. We shall also find in such electric currents an explanation of sun-spots. It is not meant that a full knowledge can be obtained of their minute constitution, nor is it necessary; but the equatorial belt of six degrees, nearly free from sun-spots, we can readily understand to be caused—since sun-spots are depressions in the photosphere down to the deeper and denser cloud strata beneath—by the equatorial piling up of the sun's atmosphere by its rotation. Any point on the sun's equator travels at four times the rotational velocity of one on the earth's equator, but the sun's attraction of gravity is twenty-seven and one-tenth times that of the earth, so that the piling up of an atmosphere of hydrogen would be considerable, and such depressions would not ordinarily exist there. Similarly, near the sun's poles we should find a gradual darkening, as is the case; but from five degrees to thirty degrees latitude, the sun, in its rotation, by reason of the inclination of its axis, passes at every point directly beneath the planets, or within their area of control, and here we find the solar spots in their greatest number, size, and intensity. These sun-spots cross the face of the sun in about fifteen days, and vary in development from year to year, having a cycle of 11.11 years from maximum to maximum. They also have a long cycle of about fifty-six years. (See article "The Sun," in Appleton's Cyclopædia.) "Wolf, in 1859, presented a formula by which the frequency of spots

108   SOURCE AND MODE OF SOLAR ENERGY.

is connected with the motions of the four bodies, Venus, the earth, Jupiter, and Saturn. Professor Loomis, of Yale College, has since advocated a theory (suggested by the present writer [Proctor] in 1865, in 'Saturn and his System,' page 168, note) that the long cycle of fifty-six years is related to the successive conjunctions of Saturn and Jupiter. But the association is as yet very far from being demonstrated, to say the least." Should such fact

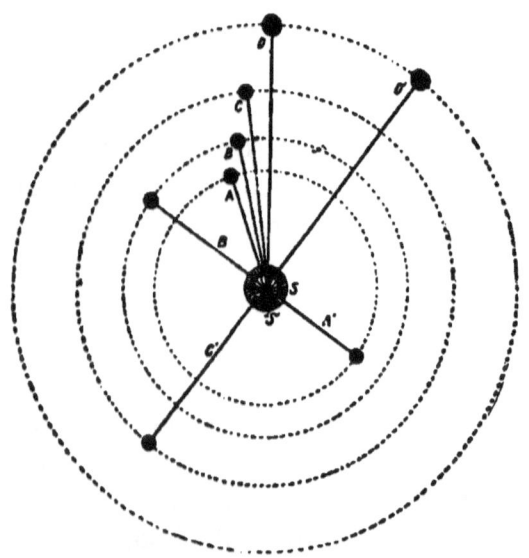

Position of planets with reference to the generation of sun-spots.—S, the sun; S', axis of sun's rotation inclined 7° to plane of planetary rotation; A B, C, D, maximum intensity of planetary action; A', B', C', D', minimum intensity of same.

be established, an explanation for it will be found in the direct impact of the condensed electric currents from several planets approaching conjunction, and raising a portion of the sun's atmosphere suddenly to a higher temperature and volatilizing an abnormal proportion of the semi-vaporous metallic core beneath. This would form an upburst piling the intensely heated faculæ up on the sides and revealing the relatively darker masses of cloud beneath, the cooler supernatant hydrogen pouring in from the upper layers to fill the returning void. This is precisely what is seen in such spots and their surrounding disturbances. In the article "The Sun," above quoted, we read, "Mr. Huggins has found that several of the absorption bands belonging to the solar spectrum are wider in the spectrum of a spot, a circumstance indicative of increased absorption so far as the vapors corresponding to such lines are concerned. . . . Near the great spots or groups of spots there are often seen streaks more luminous than the neighboring surface, called *faculæ*. They are oftenest seen towards the borders of the disk." This writer also describes "luminous bridges across spots which sink into the vortex and are replaced by others of the numberless cloud-like forms from one hundred to one thousand miles in diameter, the brilliancy of which so greatly exceeds that of the intervening spaces that they must be recognized as the principal radiators of the solar light and heat." The apparent retardation of the spots most distant from the sun's equator may also be partially, at least, explained by planetary currents of electricity, as the equatorial atmosphere is deeper and more likely to carry forward such vortices when formed, while the planets act more directly on the sun's mass beneath their direct influence.

Let us consider this retardation of sun-spots somewhat more in detail. Take, for example, the case of a large planet at such orbital position that its direct line of electrical impact will penetrate the photosphere at (say) seven degrees north solar latitude, which is about fifty-two thousand miles from his equator. During its annual revolution this planet will traverse, with its line of energy, every point of the sun's surface down to seven degrees south latitude and back again to its initial point, thus tracing a close spiral around the sun for fourteen degrees, or about one hundred and four thousand miles in width. The centrifugal force of the solar rotation piles up the photosphere and the chromosphere around the sun's equator, precisely as our atmosphere is piled up around our own equator. If the planet be a large one (for distance has but little to do with these electrical currents at planetary distances, in which they differ entirely from light, heat, and gravity), or if there be two planets nearly in conjunction, the body of the chromosphere and the surface of the photosphere will gradually become highly heated, for currents of electricity, of themselves, do not directly heat the solar core any more than a like current heats the under carbon of an arc lamp, the high temperature in both cases being altogether due to the incandescent heat of the interposed arc or envelope. Faculæ of intense brightness will then appear upon the photosphere, and these will be driven forward and also outward in the direction of the higher latitudes, producing an oblique forward movement from difference of rotational speed at different portions of the sun's surface. Similar phenomena are constantly observed on the surface of the earth in the generation and behavior of cyclones and other atmospheric disturbances. They may be compared to the wake of a vessel anchored in a strong tide-way. These faculæ will slowly raise the temperature of the surface of the sun's core beneath to the point of eruptive volatilization, and particularly so if the planet is receding from, instead of advancing towards, the solar equator. At some point in advance of the line of planetary energy an eruption of volatilized metals will suddenly occur, first thrusting up a vast area of the photosphere and then bursting it asunder, which will drive these ruptured masses with enormous speed forward and obliquely outward from the equator. Such faculæ (see Proctor's "Light Science") sometimes reach a velocity of seven thousand miles per minute, while the sun's rotational movement at the equator is less than seventy miles per minute. This sudden eruption will be almost immediately succeeded by great expansion and consequent fall of temperature, so that within a few hours the heavy volatile metals begin to condense and rapidly recede into their crater, and the faculæ in front and at the sides will now stream inward to occupy this vacuum with constantly accelerated velocity, pouring over the edges like the rush of waters at the Falls of Niagara. As they sweep downward over the inner rim of the funnel, these streams of faculæ will glow with increased whiteness, and appear to

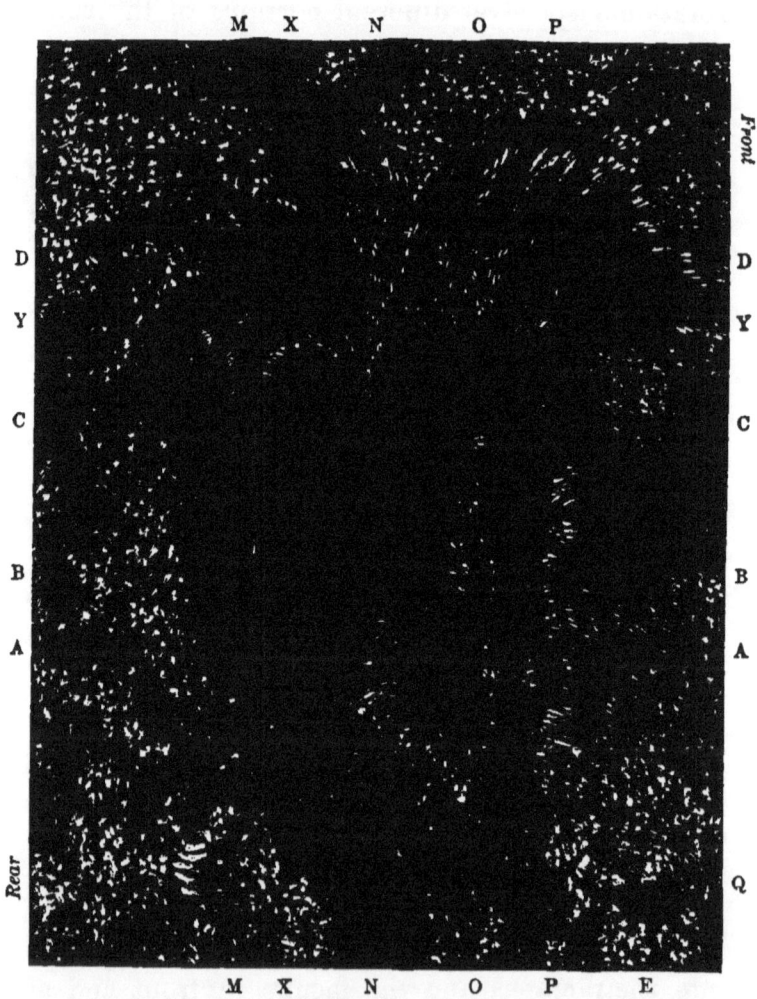

Analysis of a typical sun-spot. Intersections of lines drawn between AA and MM, CC and MM, show state of active eruption; DD, inflowing faculæ pouring downward over the rim; PP, the same; OO and BB a floating bridge, partially completed, supported by the uprush, and along the line NN torn asunder, and upward into plumes and sprays. The general surface shows the mottlings and faculæ. The partial formation of a loop is shown at XX, YY. The line EQ represents the sun's equator; from *rear* to *front*, the direction of solar rotation. The line of planetary impact is in rear.

be sharply cut off at their inner ends; but this is only apparently so, and is due to the position of the observer, who looks almost directly downward upon these descending streams. It is for the same reason that the faculæ appear more brilliant when near the borders of the solar disk (see page 109). Any good view of a sun-spot when analyzed will show the streams of faculæ thus pouring inward, and they are among the most peculiar and conspicuous phenomena to be observed. The drawings of Professor Langley, reproduced in the *Popular Science Monthly* for September, 1874, and July, 1885, are particularly striking in their illustration of these effects, though their significance and interpretation were not then at hand.

But while these heavy metallic vapors so rapidly condense and subside in the forward or initial portion of the sun-spot under observation, new depths of intensely-heated faculæ are generated behind, and these operate with renewed energy upon the fresh surface of the solar core in rear of the original seat of eruption; so that each sun-spot, while in an active state, will exhibit two entirely distinct aspects, the forward portion of the crater in a state of rapid condensation and subsidence of the recently erupted metallic vapors, and with inflowing streams of incandescent hydrogen from the front and sides, and the rear portion of the crater up to its rearward wall, and even streaming forth from beneath it, in a state of violent eruption. The large volcanic craters of the Hawaiian Islands exhibit similar partial eruptions and subsidences progressing simultaneously in the same depths. The sudden formation of the great incandescent loops and plumes to which Professor Langley calls especial attention, and which have hitherto been so perplexing, can now be readily understood and ex-

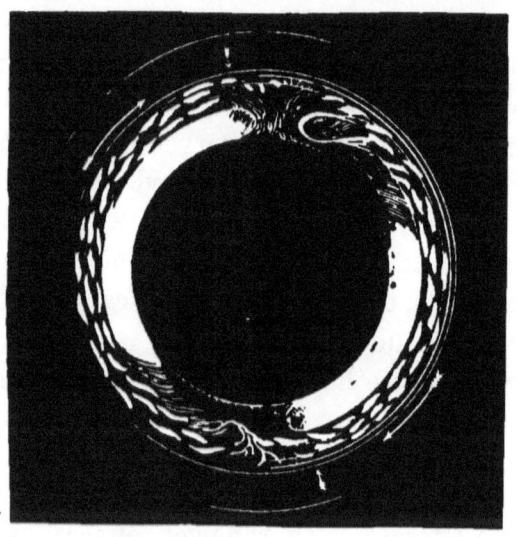

Retardation of sun-spots by continuous development to the rear, and recession in front, as the sun rotates on its axis. The short arrows represent lines of planetary energy; the long arrows show the direction of the sun's rotation.

The dark inner disk represents the solar core, the white circle the photosphere, the mottled area the chromosphere and faculæ, and the dark outer ring the corona. Loops and tufted sprays are shown, caused by inflowing faculæ in front, caught by the uprush of active portions of the sun-spot towards rear.

plained. If one of these inflowing streams be carried partially down into and across the crater, and then caught, in its advance, by the uprush in the central or rear portions of the cavity, it will be at once swept upward alongside the ascending eruption, and either scattered at its forward extremity into sprays and plumes, or else thrown forward bodily in the form of a more or less complete loop. In a sun-spot fifty thousand miles in diameter, such a loop, having a long diameter of twenty thousand miles, if we give a speed to the faculæ of seven thousand miles per minute, would be formed in about seven minutes, during which the sun-spot would itself have advanced less than five hundred miles across the face of the sun. The luminous bridges which form so suddenly across portions of the crater may be explained in a similar manner: they are streams of faculæ floated on the nearly balanced uprush of metallic vapors from beneath.

It will thus be seen that a sun-spot is not merely a fixed eruption, like a volcano, but rather a continuous series of eruptions, like a line of activity following, for example, the great terrestrial volcanic curve which extends up the western coast of America, across the Pacific Ocean and Asia, and into Central and Southern Europe, for during its progression its scene of action is constantly being shifted to the rear; it is like a furrow cut by a plough, in which the upturned sod is constantly falling in at one end of the furrow while the plough is cutting a new furrow at the other, except that in this case the plough is relatively fixed overhead, and the field itself passes along beneath it. Consequently, the center of activity of a sun-spot is only in its rear portions, generally considered, and the whole sun-spot is gradually retreating, by successive filling up in front and opening out behind, farther and farther to the rear,—that is to say, to the east,—so that retardation relatively to the rotational advance of the photosphere necessarily ensues.

But when the sun-spot is developed upon or near the equatorial line this retardation is not so considerable, for the deeper layers of the photosphere in those regions are slower to act and require greater energy to affect them, so that all except deep and violent eruptions fail to show themselves at the surface at all, and the heated faculæ are carried directly forward along the surface of the equatorial swell, so that the center of activity is driven forward more rapidly than in the higher latitudes, and the rate of progression is more nearly coincident with that of the photosphere. But if these facts are correctly stated and explained, we may have to revise our calculations of the sun's rotational period, for retardation to some extent must occur in all cases, if in any.

A sun-spot, we thus perceive, is an elongated wave or ridge of eruption along the rotational direction of the sun's body. Why, then, it may be asked, is not this line of eruption continuous entirely around the sun? For the same reason, it may be answered, that our own cyclones are not continuous, though caused substantially in the same manner, and that volcanic eruptions only occur at long intervals, though the forces at work are continuous. Lowering of temperature follows swiftly after eruption, and as the deeper structures of the solar nucleus become gradually affected, instead of volatilization of the outer layers of the surface, we will have diffused gaseous expansion of large portions, and finally of the entire solar mass, which cannot as a whole be volatilized by any conceivable planetary energy. We see these operations exemplified in heating a bar of copper in a Bunsen flame; the latter first turns green from surface volatilization of the copper, but as the heat is communicated to the deeper structures the green flame disappears, and the whole additional heat goes to raise the temperature of the mass.

These processes in the sun are thus seen to be self-compensatory in their nature. They are the means provided to distribute the restricted areas of abnormally heated photosphere over the solar surface, and finally to cause the absorption of the whole excess of heat in the sun's central mass. The balance is so evenly maintained, however, that, were all the planets equally distributed with reference to the sun's surface, such sun-spots would be the exception and not the rule, and their distribution would be equal and constant; but, as the planets continually change their positions with reference to the sun and to each other, only by some such provision of nature could the internal structure of the sun be maintained without serious derangement, or, indeed, final disruption. So nature distributes her stores of heat upon the earth. These beautiful self-compensations we shall find suddenly appearing, as we advance, in all parts of the field of astronomical research.

It may seem like temerity to advance statements so positive and specific as to the cause, constitution, and progression of sun-spots, in the absence of any considerable accumulation of observations to sustain them, but the few examples which we have noted are in accordance with these views, and when attention is once called to the basic principles on which they depend, observations will doubtless be made in abundance to prove or disprove what has been here stated. The mere fact of a differential rate of advance among sun-spots, as they pass across the solar face, of itself demonstrates that the active causes of these phenomena must be extra-solar, and if so, their only possible dynamic source must be looked for in the planets, and the remaining conclusions will of necessity follow as a corollary. We may even, by merely examining an accurate drawing of a sun-spot, determine its position and direction upon the solar sphere from which it was delineated by its lines of active eruption and influx of faculæ, and also whether it be a new spot or one which has passed entirely beyond its active stage and is about to finally disappear.

As for the faculæ which striate the photosphere, the mottlings and so-called "willow-leaves," any one who will quietly gaze downward upon the turbid surface of the Mississippi or other similar river, in mid-channel, will see plenty of such faculæ: the river is full of them. The heavier, intermingled clay, slowly subsiding, is caught up in the turmoil beneath the surface and swept upward in elongated ovals and eddies, the larger swells nearly colorless, and others of all shades of ochre and yellow, and the whole as richly mottled, sometimes, as the variegated pattern of a Persian carpet. If we substitute for the subsiding clay the rapidly sinking heavy metallic vapors, and enlarge the scale from the dimensions of the river to those of the sun, we will have the mottled solar surface with its kaleidoscopic changes, the so-called "willow-leaves," and the faculæ in all their glory. A careful study of the sun will show most clearly that only in some such explanation as the present view affords can a rational basis for its varied phenomena be found.

If the sun's equator were coincident with the plane of the planetary orbits, it is obvious that all the planetary energies would be directed, whatever the position of the planets around the sun, immediately upon this equatorial great circle, and that, at each revolution upon his axis, corresponding nearly to our calendar month, the same part of his sphere would be exposed to these direct currents, so that the intensity would be, in its aggregate, nearly a constant quantity. But, by reason of the sun's axial inclination of seven degrees to the plane of the planetary orbits, a far more complex and important condition of affairs ensues. It will be seen at once that the plane of the planetary orbits intersects the sun's equator at opposite sides, and that, from a minimum of nothing, this line reaches a maximum, twice in each circumference, of seven degrees, one north and the other south of the equator, and that this arc of fourteen degrees, thus traversed by every planet in its orbital rotation around the sun, measures more than one hundred thousand miles from north to south upon the solar surface, nearly one-half the distance which separates the

120   SOURCE AND MODE OF SOLAR ENERGY.

earth from the moon. If all the planets were in conjunction or nearly so, on one side of the sun, for example, and in the vertical plane of the sun's axis, they would continue to deliver their electrical

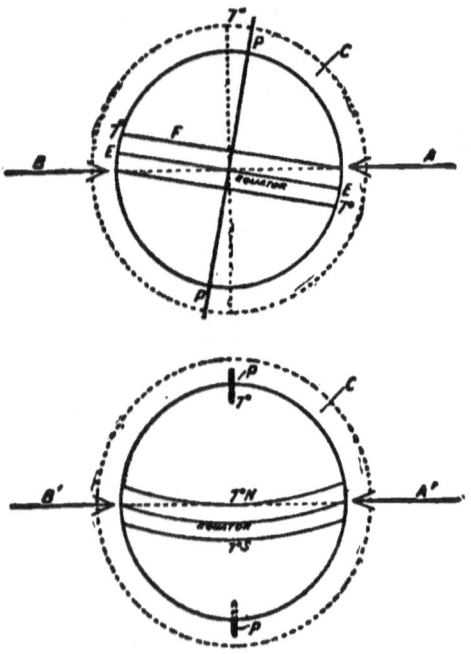

Illustrating complex lines of planetary electrical energy produced by inclination of sun's axis.—A B, A' B', plane of planetary orbits.

Upper figure shows sun's axis inclined laterally; lower figure, from front to rear, and at right angles to former.

C, chromosphere; E E, solar equator; A B, A' B', lines of planetary electric currents; F, latitude covered by vertical position of planets, 14° in width; P P, sun's axis.

currents with their greatest intensity upon a single point of his surface fifty-two thousand miles north of his equator, while the opposite point, one hundred and four thousand miles distant, would be unaffected by any direct currents at all. Conversely, if in conjunction on the opposite side of the sun, they would continue to deliver these currents upon a corresponding point fifty-two thousand miles south of the equator; but if in conjunction in the vertical plane transverse to the sun's axial inclination, these currents on either side of the sun would be delivered directly upon the solar equator. The importance of this will be understood when it is considered that for many of our years such planets as Jupiter and Saturn must continue to direct their currents upon a very slowly changing point of the sun's surface, by reason of their vast annual rotational period, while with the earth and the interior planets these various points are struck with ever-increasing rapidity as the year decreases in length with the different planets, the earth, Venus, and Mercury. There is a solar equinoctial, so to speak, just as there is a terrestrial equinoctial in which the sun crosses the line twice each year, and the meteorological disturbances faintly shown on the earth at such times are vastly increased on the sun, and rendered far more complex by the interaction of many planets upon the sun, instead of a single sun upon each planet. While our equinoctial has to do with gravity and light and heat, and probably magnetism, the solar equinoctial deals with the vast electrical streams which feed its fires and set it boiling with furious energy, first at one point, then at another, until the increment has been absorbed and adjusted, and thus equalized throughout his mass. What a new interest this must arouse in our study of sun-spots, faculæ, prominences, sun-storms, and the vast panorama of solar action hung up before our astonished eyes! A new world here awaits its Columbus.

But not only the planets thus gather, so to speak, electricity for the sun's support from space; the moon also must do its part, as it rotates in the same manner, subject to the sun, and has its own motion through space. But an examination of the moon shows no atmosphere and no aqueous matter visible to us, and also the singular fact that it constantly presents one side only to the earth. R. Kalley Miller, in his "Romance of Astronomy," article "The Moon," says, "After an elaborate analysis, Professor Hausen, of Gotha, found that it could be accounted for only by supposing that the side of the moon nearest us was lighter than the other, and hence that its center of gravity was not at its center of figure, but considerably nearer the side of it which is always turned away from us. He calculates the distance between these centers to be nearly thirty-five miles, evidently a most important eccentricity, when we remember that the radius of the moon is little over a thousand miles. It must have been produced by some great internal convulsion after the moon assumed its solid state; but the forces required to produce this disruption are less than might at first sight appear necessary, owing to the fact that the force of gravitation and the weight of matter are six times less at the moon than with us." Those who are fond of the so-called "Argument of Design" will be gratified to learn that, if the moon had a rotation upon its own axis similar to that of the earth, all life—past, present or future—would have been impossible on that satellite or planet; and that, on the contrary,—provided she always turns the same side of her surface to the earth,—it is quite possible that air, water, and life may exist, or may have existed, on the opposite side of the moon, but not otherwise. In fact, air and water must now exist on the opposite side; and, since her whole supply will thus be condensed upon half her surface or less, even with her small force of gravity, it may be quite sufficient in quantity and density for the support of animal, vegetable, or even human life. By reason of this difference in the lunar center of gravity, the side presented to the earth in physical position is similar to the summit of a mountain upon the earth's surface two hundred miles high, and surely we would not expect to find much air or water or life at that altitude. But the opposite side would resemble a champagne country at the foot of this enormous mountain, and might be well fitted for human existence. Now, we know that similar electricities repel each other, and air or gases charged with similar electricities are equally self-repellent. Professor Tyndall, in his "Lessons in Electricity," says, "The electricity escaping from a point or flame into the air renders the air self-repulsive. The consequence is, that when the hand is placed over a point mounted on the prime conductor of a good machine, a cold blast is distinctly felt. . . . The blast is called the 'electric wind.' Wilson moved bodies by its action; Faraday caused it to depress

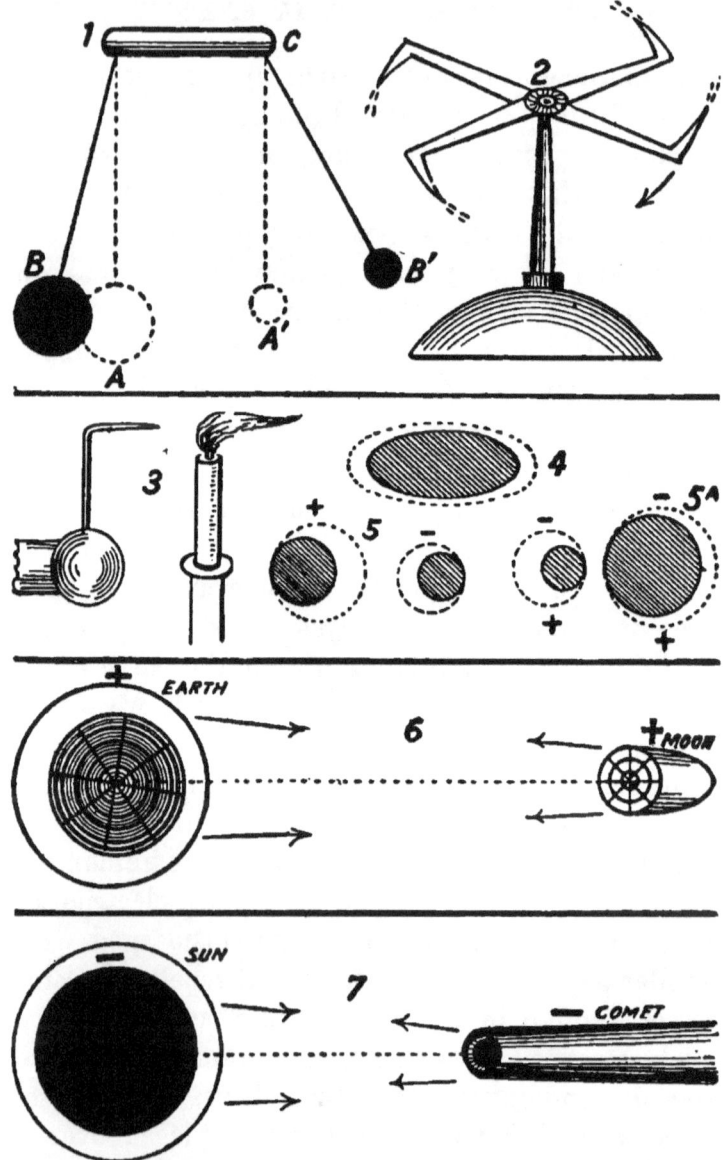

Fig. 1, mutual repulsion of similarly electrified pith-balls; 2, the electrical windmill, atmospheric repulsion; 3, repulsion of a flame by electricity; 4, electrical distribution around an oval conductor; 5, mutual attraction of opposite electricities; 5a, mutual repulsion of similar electricities; 6, mutual repulsion of electrospheres of earth and moon; 7, mutual repulsion of electrospheres of sun and comet.

the surface of a liquid; Hamilton employed the reaction of the electric wind to make pointed wires rotate. The wind was also found to promote evaporation."

While electrical repulsion is doubtless analogous to, and correlative with, the attraction of gravitation, this force, and even gravity itself, has been sometimes interpreted as derived from the mutually interacting molecules of space itself. We may even learn somewhat of how such repulsions of similar and attractions of opposite electrospheres might occur. We constantly speak of positive and negative electricity as though these were different fluids, but such expressions are employed only in the same manner as the analogous terms, heat and cold. We know, of course, that cold is the relative absence of heat, the dividing line being not a fixed, but a constantly changing one, so that one body is cold to another by reason of relative, and not absolute, deprivation of heat. It is well known, however, that cold, which is purely a negative state, manifests the same apparent radiant energy as heat. A vessel near an iceberg is exposed to a wave of cold, precisely as of heat from a heated body at the same distance. This, of course, is due to abstraction and not to increment. All space being occupied by attenuated matter in a state of unstable electrical equilibrium, as we say, which simply means a condition ready to be raised or lowered in tension by absorption from or into outside media, all concrete bodies floating in that space must have an electrical potential either equal to, or higher, or else lower than that of their surrounding space. A solitary body in space, if we can conceive of such, in either a higher or lower state of electrical tension, would be drawn upon from all sides to equalize the distribution and restore the general average. But if two bodies occupy the same field, and are widely different from each other in electrical potential, one higher and the other lower than that of space, this distribution will be towards each other, and must be manifested by mutual attraction. But if, on the contrary, these two bodies are both equally higher or lower than the spatial average, they have nothing to give to each other, but have this difference to give to or receive only from outer space, and hence they will be drawn apart or, as we say, mutually repelled. The case is similar to what we see in the case of bodies of water at various levels. Suppose there be a lake of a fixed level, and communicating with it and with each other, by open channels, two ponds of water occupying an island in the middle of the lake. If one of these ponds be higher in level and the other lower than the lake, their waters will rapidly converge, the higher flowing into the lower; but if both are at the same level, and higher than the lake, they will flow apart into the lake. Or, if both are at the same level, and lower than the lake, the water of the latter will equally flow from outside into both ponds, and their waters will still be held separate from each other. The analogies of these various levels may be pursued to any desired extent, as electrical ten-

THE SOURCE OF SOLAR ENERGY. 127

sions find their most exact analogies in the pressures of bodies of water at different levels and of different quantities, and these analogies are those most constantly used in the interpretation of such electrical phenomena.

The great electrical activity of the electrospheres of the earth and moon, while they discharge their tremendous currents directly into the sun, at the same time must cause their similarly electrified atmospheres to mutually repel each other, while gravity continues to operate to maintain the earth and moon at their fixed distances from each other, and to retain their gaseous envelopes around their own bodies. The result must be that these similarly electrified atmospheres repel each other with a force proportioned to their masses of atmosphere and the intensity of the electricities of each. The moon's axial rotation being completed but once in twenty-eight days, and that of the earth once in each day, and the moon's mass and volume being so much less than those of the earth, whatever of electrified air or moisture she may have (and she *must* have both, proportionate to her attributes) would have been driven as by a cyclone to the opposite side of the moon and there retained. Now, with an atmosphere and water only on one side of the moon, and that the side opposite the earth, it is obvious that a rotation on her axis at all resembling that of the earth would carry every part of her surface, at each complete rotation, from a region of air and moisture into one deprived of both, and in such a condition she would of necessity be deprived of both life and its possibility; hence, as the laws of nature compel the lunar atmosphere and moisture to reside permanently on the side always opposite the earth, a co-ordinate arrest of the moon's axial motion with reference to the earth could alone compensate for such a state of things, and, curiously enough, we find as a solitary exception, compared with the planets, that such is the case. The moon unquestionably has both atmosphere and water on its opposite side. In his recent work, "In the High Heavens," Professor Ball reviews the physical conditions of the other planets as possible abodes of life. He pronounces against the moon because night and day would each be a fortnight in length; but this is surely no objection, for even in Norway and Greenland such nights and days are not uncommon at different seasons, and thousands of human beings, even as at present constituted on earth, spend their lives there in content and happiness. That the moon also would be terribly scorched by the long day and frozen by the long night does not necessarily follow, for the atmosphere of Mars, that author says, "to a large extent mitigates the fierceness with which the sun's rays would beat down on the globe if it were devoid of such protection." As the moon's opposite face must have a double quota both of atmosphere and clouds, the difficulty will be correspondingly less than on Mars; and as for the "lightness" of bodies on the moon, they would probably get along quite as well as mosquitoes and like "birds of prey" in the marshes along our coasts. The author refers constantly to *our* bodies; for example, "Could *we* live on a planet like Neptune?" No, we could not; we would be dead before we got there. Nor could *we* live in the bark of a tree, or at the bottom of the ocean, or in a globule of serum; but living beings are found there nevertheless. The principle is that wherever life is possible there we may expect to find life; and surely life is, or has been, or will be possible, not only on the moon, so far as our knowledge of physical conditions can go, but also on some of the other planets. Of course each planet has its life stage, but this applies not only to the earth, but to all the other planets as well, and not only to the planets of our own system, but to those of all other solar systems. Each has had, or will have, its stage in which life is possible, and these planets may be like human habitations, in which whole races at times migrate from one home to another. There is no conceivable reason why this may not be the general law of creation, and every analogy leads us to believe that it is so.

It has been recently announced that, from telescopic observations, the atmosphere of Mars must be at least as attenuated as that among the highest mountainous regions of the earth, if this planet has any atmosphere at all. That it must be far less dense than that of the earth at sea-level is obvious, for the mass and volume of Mars are very much less than those of our own planet; but that Mars is devoid of a gaseous envelope or atmosphere is contrary to what we know of all sidereal physics. The sun, the fixed stars, the comets, the nebulæ, and even the meteorolithic fragments which fall upon the earth, all show the same elementary chemical constitution as the earth itself, and we cannot believe that Mars alone is differently constituted from every other body we have been able to examine. We have direct evidence, on this planet, of polar snows and their melting away under the sun's heat; we see the apparent areas of sea and land; it has its moons as the earth has hers, and exhibits all the characteristic phenomena of the earth and other planets. All sidereal bodies that we know of, except, perhaps, our moon, which exception we have fully accounted for, are found to be surrounded by gaseous envelopes or atmospheres of some sort. The sun, the fixed stars, the nuclei of comets, the condensing nebulæ, the planets Jupiter and the earth, which are those under our most direct observation, and even the meteorites, when examined, reveal the presence of many times their own volumes of independent atmospheric gases; and whatever may be the theory of the origin or development of Mars, it must have been subjected to the same influences, the same environment, and the same processes of creation as those of our solar system generally; and that this body alone should possess no gaseous envelope—for the denial of atmosphere denies, at the same time, the presence of any or all surrounding gases—is quite incredible. Only the most positive, direct, and long-continued proofs of such fact could be accepted, and even then the history of all scientific progress shows that what are believed to be facts themselves fluctuate like fancies till, by their accumulated force, they solidify into universally accepted demonstration. The fact, moreover, that the atmospheres of the smaller planets are more attenuated than our own and those of the larger ones denser has no bearing, in itself, on the probability of the existence of life on these other planets, for in our own atmosphere oxygen, which is the efficient element, is diluted with four times its quantity of inert nitrogen. These proportions doubtless vary largely in other atmospheres, so that the oxygen may be much richer in some and far poorer, relatively, in others. The mere fact that the presence of nitrogen, probably, and aqueous vapor, certainly, depends on the gravity of the mass of each planet, while the oxygen is due to electrolytic decomposition induced by the combined volume, mass, and rotation, and other causes,—such as the axial inclination of such planets, for example, —renders these variations in the constitution of planetary atmospheres a certainty. As Mars has a diameter much more than one-half that of the earth, and a diurnal rotational period nearly the same, while his mass, which controls the action of gravity, is only about one-ninth that of the earth (see Appleton's Cyclopædia), it is obvious that his oxygen-gathering power, compared with that for accumulating nitrogen and aqueous vapor, is much higher than that of the earth, and we should expect to find there an attenuated atmosphere very rich in oxygen, and with a relatively smaller proportion of aqueous vapor, or even water, on his surface. Such seem to be the facts as far as observed.

In operating an electric machine the strength of the current is directly proportionate to the speed of rotation,—that is to say, to the velocity of the generating surface; for example, of the Wimshurst induction machine it is stated (page 63, "Electricity in the Service of Man"), "These four-and-one-half inch discharges take place *in regular succession at every two and a half turns* of the handle." It is also a well-established law of electrolysis that "The amount of decomposition effected by the current is in proportion to the current strength." Professor Ferguson ("Electricity," page 225) says of the voltameter, an instrument devised by Faraday, and used for testing the strength of currents by the proportionate decomposition of acidulated water, "Mixed gases rise into the tube, and *the quantity of gas given off in a given time measures the strength of the current.*" Roughly estimating the diameter of Mars at five-eighths, the surface velocity at three-fifths, and the mass at one-ninth those of the earth, this planet should have an atmosphere containing about sixty per cent. of oxygen and forty of nitrogen, with a barometric pressure at sea-level of about six and one-half inches of mercury. This would be an excellent atmosphere,—about equal in its quota of oxygen for each respiration to that of the higher areas of Persia, a great country for roses. The aqueous vapors lying low and near the surface would serve as a vaporous screen to concentrate and retain the sun's heat and retard radiation from that planet. Nothing in particular seems to be the matter with Mars.

THE SOURCE OF SOLAR ENERGY. 133

On the contrary, the mass of Jupiter is so great, and his attraction of gravity so powerful, that it is only by his exceedingly rapid diurnal rotation (once in less than ten hours) that it is possible for him to accumulate any effective percentage of oxygen at all. But there is certainly plenty of water there.

We may approximately compute, in general terms, the proportion of oxygen in the atmospheres of the other planets in the same way. Neptune, it is true, is so far distant from the sun that the solar orb only "appears about the same magnitude as Venus when at its greatest brilliancy, as viewed from the earth," but we must not forget that "the *intensity* of the sun's light would be more than ten thousand times greater than that of Venus" (Professor Dunkin, in "The Midnight Sky"). Unless the moon gathers a portion of the earth's oxygen (the planetary satellites, like Saturn's rings, thus constituting in their rotations a constituent part of the planets themselves), the percentage of this gas in her atmosphere must be exceedingly small, for her axial rotation has a period of a whole lunar month, being the same as that of her revolution around the earth as a center.

The absence of apparent atmosphere and moisture from the *visible* lunar surface has already been mentioned and explained. The means by which this fact has been approximately determined are described by Professor Dunkin, in "The Midnight Sky," as follows: "Among the many proofs of the non-existence of a lunar atmosphere, it may be mentioned that no water can be seen; at least there is not a sufficient quantity in any one spot so as to be visible from the earth. Again, there are no clouds; for if there were, we should immediately discover them by the variable light and shade which they would produce. But one great proof of the absence of any large amount of vapor being suspended over the lunar surface is the sudden extinction of a star when occulted by the moon. The author has been a constant observer of these phenomena, and, though his experience is of long standing, he has never observed an occultation of a star or planet, *especially at the unilluminated edge of a young moon*, without having his conviction confirmed that there is no appreciable lunar atmosphere. . . . Professor Challis has subjected the results of a large number of these observations to a severe mathematical test, but he has not been able to discover the slightest trace of any effect produced by a lunar atmosphere."

In Appleton's Cyclopædia, article "The Moon," it is stated that "Schröter (about 1800) claimed to have discovered indications of vegetation on the surface of the moon. These consist of certain traces of a greenish tint which appear and reappear periodically; much as the white spots covering the polar regions of Mars. . . . As we are able, under the most favorable conditions, to use upon the moon telescopic powers which have the effect of bringing the satellite to within one hundred and fifty to one hundred and twenty miles of us, we should doubtless notice any such marked changes on her surface as the passage of the seasons produces, for example, on our own globe." Very recently (August 12, 1894), it has been stated, Professor Gathmann has observed a peculiar green spot about forty by seventy miles in area near the crater of Tycho Brahe, "on the *northwestern edge* of the satellite's upper limb," which had disappeared twenty-two hours afterwards.

We understand, of course, that the moon's librations, by the variation of position of the lunar body, enable us to see, at times, around the edge of this satellite somewhat, so that, instead of observing only one-half, we can in this way see nearly six-tenths of her surface, but not at the same time, of course. When the moon is dark it occupies a position between the earth and the sun, and only its opposite face is illuminated. In this position the attraction of solar gravity and the attraction of the electrically opposite solar electrosphere both accumulate their forces upon the moon's atmosphere in the same line as the repulsion of the earth's similar electricity, so that the lunar moisture and atmosphere are, at this part of her subordinate orbit, most powerfully forced away from the direction of the earth. As the moon now proceeds towards her first quarter, the terrestrial repulsion drives her atmosphere radially outward, while solar gravity and electrical attraction tend to hold it in the direction of the sun. The result will be an electrospheric libration, so to speak, and the moon's atmosphere and moisture will be carried around towards its illuminated face and, to some extent, will overlap the area of terrestrial repulsion. But as the moon advances this will gradually diminish, soon cease, and finally be reversed as it again approaches darkness. We can now understand why the green surface, if it really was due to vegetation, appeared along the *lunar margin* at the time described above, and also that the observation of planetary occultations " at the unilluminated edge of the young moon" was the very worst part of the moon and its orbit in which to look for air or moisture; as the sun's influence is then directly *away from* the unilluminated surface of the moon, and his "pull" would have, in fact, still further denuded the very portion most persistently examined, and where this absence of atmosphere was *especially* noted.

When considering the transference of energy from the peripheral regions of the solar system to the center, its conversion there into a new form of molecular force, and its subsequent distribution, we find a curious and instructive parallel in the action of the reflex nervous system of animal life. This system is one in which the brain or other conscious center of nerve-energy takes no part. Tickle the foot of a child, for example, and its whole muscular system is thrown into uncontrollable convulsions of laughter. Here an exciting contact with the terminal filaments of the afferent or sensory nerves is rapidly carried into the local nerve-center of this part of the system,—that is, the sensory column of the spinal cord. This center of ganglionic nerve-matter lies directly against the corresponding

THE SOURCE OF SOLAR ENERGY. 137

motor mass, both freely communicating with each other. The sensory current passing into its central ganglion undergoes some peculiar change of character, probably one of intensification, such as is observed in the action of the condenser of an electrical machine, through which sensory ganglion, thus raised in potential, it passes to the motor ganglion adjacent, where it is instantly transformed into an entirely different form of energy. The sensory character has now entirely disappeared, and it has been converted into and is flashed forth as motor energy to the different muscles of the body, which are immediately contracted, the violent molecular motion of the fibres being at once converted into muscular motion in mass. The changes are entirely analogous to those we see in the different conversions of energy in our solar system. Considering the surface of the body as a planetary electrosphere, it is acted upon by excitation from without; currents of energy are engendered, which are at once transmitted to the sensory ganglion, corresponding to the hydrogen atmosphere or electrosphere of the sun; intensification of action here ensues, the current passing through this ganglion or atmosphere into the solar body itself, which corresponds to the motor ganglion; both ganglia are now highly excited; the electrical force is converted into the radiant molecular motor energy of heat and light in the sun and muscular excitement in the body, and these are flashed forth and find scope for their action within the body of the subject or upon the surface of the planets, which lie, like the muscular structure of the body, within the genetic electrosphere where, acted upon from without and by agencies entirely external, moving contact has induced the primary molecular action, which was then instantaneously transferred to the center, there converted into another form, that of motor energy, and thence sent forth to produce action in the muscles of the body in the one case, and in the other upon the planetary bodies and their satellites and other structures which occupy surrounding space.

CHAPTER V.

THE DISTRIBUTION AND CONSERVATION OF SOLAR ENERGY.

WHAT, then, becomes of the light and heat flashed forth with eternal energy from the fiery waves of the sun's incandescent atmosphere? Professor Ball ("In the High Heavens") says, "Much of what has been said with regard to light may be repeated with regard to heat. We know that radiant heat consists of ethereal undulations of the same character as the waves of light. Hence we see that the heat or the light radiated from a glowing gas is mainly provided at the expense of the energy possessed by the molecules in virtue of their internal oscillations." Conversely, of course, the ethereal undulations thus induced by high molecular motion in the heated gas or vapor must disappear in so-called absorption or transference by contact with other molecules, themselves devoid of such specific internal oscillations. The heat motion then disappears as heat by its conversion into work, just as the motion of a belt in a mill disappears in the work of the machine which it drives. One two-hundred-and-thirty-two-millionth part of the radiant solar energy, we know, is caught by the flying planets of our system in the forms of heat and light, adapted to sustain life and its continued potentiality, and we know that this solar energy is the sole source of all the development and maintenance of the planets as the possible abodes of organic life, past, present or future.

But what of the vast total, of which we consume so minute a fraction? It is true that, in addition to the planets, space is occupied by many small meteoric bodies, which manifest themselves to us as shooting stars and meteorites, but the mass of these is too trifling to be estimated. Professor Helmholtz, in his "Popular Scientific Lectures," says, "According to Alexander Herschel's estimates, each stone is, on an average, at a distance of four hundred and fifty miles from its neighbors." When these bodies enter our atmosphere by force of the earth's attraction they are heated by its atmospheric friction to incandescence, and in most cases are even volatilized before reaching the earth's surface. The vast volumes of solar heat and light, however, are poured forth from the sun indiscriminately in all directions into illimitable space, wherein all the masses of concrete matter, including the stars, are relatively far less in volume than the flying motes of the purest morning air which sparkle in the flood of light sent forth by the rising sun. Is all the rest wasted? Professor Balfour Stewart, in his work "The Conservation of Energy," says, "If this be the fate of the high-temperature energy of the universe, let us think for a moment what will happen to its visible energy. We have spoken already about a medium pervading space, the office of which appears to be to degrade and ultimately extinguish all differential motion, just as it tends to reduce and ultimately equalize all difference in temperature. Thus, the universe would ultimately become an equally heated mass, utterly worthless as far as the production of work is concerned, since such production depends upon difference of temperature."

It is obvious that the starting-point taken by the author last quoted, but which, nevertheless, is in accordance with the views now generally prevalent, is diametrically opposed to that sought to be established in this work. Professor Stewart takes the sun's inherent energy as the initial point of departure, and reasons from that as to the final consequence when all its light and heat shall have been distributed or dissipated into the attenuated medium which occupies space, and which will be thus slowly heated until all space has been raised in temperature to that of the last dying sun, when all will thenceforth remain unchanged and unchangeable, silent, dark, and dead, to all eternity. On the contrary, the purpose of the present work is to establish a directly opposite principle, based, however, on demonstrated scientific facts and not on theory, that the medium which pervades all space was originally in the same equally and universally potential state (with its molecules raised to a tension constituting an unstable equilibrium) in which, practically, Professor Stewart's argument leaves it finally, and that this universal molecular energy of position was permanently maintained by the employment of the forces which afterwards, transformed into light and heat, were shed abroad by the sun in the work of again overcoming the intermolecular tension of cohesion, and that the light and heat of the sun are merely caught up again by these same or other molecules and successively employed in the same manner, while the planetary electrospheres utilize these same forces of internal tension in the generation of electricity, which, sent to the sun, is converted into light and heat, and these are again transferred to their original source. The rotation of the planets is the grand exciting cause, and the process, in its complete cycle of development, has five stages: first, planetary generation; second, transference by currents of electricity to the sun; third, conversion into light and heat; fourth, emission; and, fifth, reabsorption and conversion again into molecular energy of position. All space is thus found to be pervaded by extremely attenuated vapors, which contain the elemental constituents out of which suns and planets are evolved under favorable circumstances of development, and, among other vapors, aqueous vapor, and that these are the agency upon which the planetary electrospheres operate in their generation of electrical currents, and which vapors, in turn, by absorption of the solar energy of radiation, again transform this energy into mutually balanced electric potential, until it is once more disengaged as electricity by the rotating planetary electrospheres, and so on in a constant circuit forever repeated. It differs from perpetual motion, however, in that the planetary rotation is the external and not the internal generative cause, since the electrical forces neither cause nor control these motions; they belong to the realm of gravity. The disassociation, moreover, is electrical and not chemical disassociation. The tensions are against cohesion and not against chemical affinity; are, in fact, similar to those which constitute our atmosphere a vast electrical reservoir; and the aqueous vapors, through all their changes, permanently remain as aqueous vapors, except those condensed portions disassociated by electrolytic action at the electrospheric poles, and which have no relation to the attenuated vapors of space, except in that the latter are their sources of supply. The process is analogous to what we see around us at all times in the atmosphere. While the process described by Professor Stewart resembles the emptying of the inherent water of a cloud, in the form of rain, into an ocean which never yields up its water again, so that, when the cloud has rained itself out, it is gone forever, the processes here sketched are like the vapors which are caught up by the heated air, carried over the thirsty lands, distributed in rain to fertilize and vivify them, then gathered in a thousand tiny rills from countless fountains, again descending to the sea and again carried up in vapor, and so on over and over in unceasing round. It is the difference between an old-fashioned flintlock musket and a modern magazine rifle, except that the magazine is always full.

This great ocean of space was primordially charged with these potential vapors; it is the constitution of space itself. We are so accustomed to consider space as empty, and that it is nothingness, the antithesis of something or anything, that it is a negation or a blank, that it requires an effort to even think of it as a fully stocked establishment with all the goods necessary for use or ornament, in the latest styles and of prime quality, only not made up, and that all our suns and worlds are merely tailoring establishments where the operatives cut and fit and make them up to order. When more goods are wanted they have to go to the store.

Is space, then, eternal, and is this constant round of energies to be eternal? If one is eternal, so is the other, and surely nothing can be more eternal than space, and we cannot conceive of any other space than this space. Out of it came all created things, and so long as the orbs rotate without retardation, so long will these interchanges go on without impairment, and that they do so rotate is the necessary corollary of the fact that they ever began to rotate. If rotation, on the contrary, was imparted by special creative power, then the same power established the laws by which they rotate, and took cognizance of resistance as well. Whatever the impulse was, it still remains; whatever caused the rotation to begin maintains it; if the cause is eternal the rotation may be eternal; and, in any case, its period must be measured by cycles of æons, to which the allotted lifetime of a dying sun—a few million years, perhaps—is but as the sunburst of a morning-glory flower to the hoary age of a mighty planet. Compared with the popular view of the sun's life-period, we may formulate the terms of an equation in which the sun's mass, compared with the realms of infinite space, is as the sun's lifetime—on a basis of contraction of his volume—to the lifetime which actually is to be. As one of the terms is practically infinite, so must be the answer to the problem. Professor Stewart says, "We cannot help believing that there is a material medium of some kind between the sun and the earth; indeed, the undulatory theory of light requires this belief." It has already been shown that the transmission of electricity also requires it, but that there must be a medium quite different from the undulatory ether. Professor Proctor ("Mysteries of Time and Space") says, "We may admit the possibility that the aqueous vapor and carbon compounds are present in stellar or interplanetary space." Again he says, "Assuming, as we well may, that space is really occupied by attenuated vapors." The same writer says further, "To this end all thoughtful study of the mechanism seems to tend (associating, perhaps, our visible universe with others, permeating it as the ether of space permeates the densest solids, and in turn with others so permeated by it); there may be that constant interchange, that perpetual harmony, of which Goethe sung:

'Balanced worlds from change defending,
While everywhere diffused is harmony unending.'"

The light and heat poured forth from the sun are, as stated, in the form of radiated energy.

They penetrate the attenuated vapors as far as vision extends, and doubtless farther, but they cannot reach the boundaries of space, for even the mind of man cannot reach those limits. Aqueous vapor absorbs heat; we know this without any demonstration, for the radiated heat of the earth is arrested by a veil of clouds, so that on cloudy nights frost will not form. So also the sun shining into water will raise its temperature, as in a glass globe, and such absorption of heat by aqueous vapors or water would be much more manifest were not a large part employed in loosening the tension of the constituent molecules, since, when thus employed, it is not manifest as sensible heat. Professor Tyndall, in "The Forms of Water," states that "The quantity of heat which would raise the temperature of a pound of water one degree would raise the temperature of a pound of iron ten degrees." Professor Stewart, in "The Conservation of Energy," says, "That peculiar motion which is imparted by heat when absorbed into a body is, therefore, one variety of molecular energy. . . . Part of the energy of absorbed heat is spent in pulling asunder the molecules of the body under the attractive force which binds them together, and thus a store of *energy of position* is laid up, which disappears again after the body is cooled.

"Heat will only be changed into work while it passes from a body of high temperature to one of low. . . . At very high temperatures it is possible that most compounds are decomposed, and the temperature at which this takes place, for any compound, has been termed its temperature of disassociation. *Heat energy is changed into electrical separation* when tourmalines and certain other crystals are heated." It may be added that it is also changed into electrical energy by the operation of all electrical machines, as molecular motions are all mutually interconvertible, and heat itself is only a mode of such motion. Of radiant energy, the same writer says, " This form of energy [radiant heat] is converted into absorbed heat whenever it falls upon an opaque substance. . . . and heats it. It is a curious question to ask what becomes of the *radiant light* from the sun that is not absorbed either by the planets of our system or by any of the stars. We can only reply to such a question that, *as far as we can judge from our present knowledge*, the radiant energy that is not absorbed must be conceived to be traversing space at the rate of one hundred and eighty-eight thousand miles a second."

We know, of course, that aqueous vapors are partially opaque to heat rays, as the radiated heat of the earth is partially arrested by such vapors in the atmosphere, but they are apparently transparent to the rays of light. But we know that this cannot be entirely true in fact, for light rays only differ from heat rays in the comparative length of their waves or impulses, while rays of light are always accompanied—when emitted by a thermally incandescent body—by a much larger number of those of heat. As a body is raised in temperature radiant dark rays first appear; these being raised higher, become visible as light, and new dark rays are radiated behind them, and this continues till after the state of highest incandescence is reached and the invisible chemical rays beyond the spectrum appear. It is like a crowd surging forth in flight from the doors of a building; as the speed of those in front increases to a run, others follow more slowly in the mass, and as these gain speed others continue to follow, while the great mass of laggards still trails along in a lengthening line to the rear. The perception of light is itself merely due to the constitution of the optic apparatus of the observer, which only takes cognizance of vibrations radiated from the middle portion of the scale, just as the ear does with sounds, and not to any actual difference in their mode of production. That heat rays and light rays are identical in constitution can be readily shown by the experiment described by Professor Tyndall in his "Forms of Water," in which an opaque screen of iodine solution in bisulphide of carbon was employed to arrest, in a beam of light, all the light waves (to which it is entirely opaque), while transmitting the dark rays. These non-luminous rays are then converged by a lens: "Let us, then, by means of our opaque solution, isolate our dark waves and converge them on the cotton. It explodes as before. . . . At the same dark focus sheets of platinum are raised to vivid redness; . . . a diamond is caused to glow like a star, being afterwards gradually dissipated." Sir William Herschel (see article "Spectrum," Appleton's Cyclopædia) says, "If we call light those rays which illuminate objects, and radiant heat those which heat bodies, it may be inquired whether light be essentially different from radiant heat. In answer to which I would suggest that we are not allowed by the rules of philosophizing to admit of two different causes to explain certain effects, if they may be accounted for by one." . . . " Tyndall, by similar experiments, found that the thermal energy of the invisible radiation of a very powerful electric light is eight times that of the visible. . . . Seebeck showed that the position of maximum heat in the spectrum changes with the nature of the prism and sometimes occurs in the red." Melconi, with prisms of alcohol and water, found it in the yellow. Athermic bands are also found in the heat-spectrum, corresponding to the Fraunhofer lines seen in the visible spectrum.

We may illustrate this successive development of more and more rapid light-waves by conceiving of a harp having musical strings of various length and thickness, but not strung up, so that, when swept by the hand, the vibrations are felt, but no musical tones are produced. If, now, all the strings are simultaneously and gradually stretched while under continuous vibration, we will first hear the hum of the lighter strings, but deep down in the scale; and as the tension gradually increases the pitch of these will rise higher and higher and be succeeded by other new tones below, until the whole register is simultaneously sounded. And if the tension be further increased, the vibrations of the upper strings will gradually grow so rapid that the ear can take no cognizance of them, corresponding to the invisible chemical rays of the spectrum, while the middle strings will be sounding loudly, and others will be slowly vibrating below the musical scale, but without sound, corresponding to the invisible heat rays. In addition to this gradual ascent of pitch along the scale, however, there is reason to believe that sympathetic vibrations are induced in the spectrum of thermal and chemical light corresponding to the over-tones in music and to those hidden rhythms which differentiate the "timbre" of one kind of musical instrument from that of another, so that a definite wave-length will not only repeat itself among adjacent molecules, but will give rise to harmonious vibrations quite different in amplitude and velocity. An example of this is found in some of the phenomena of phosphorescence and fluorescence, in which chemical rays totally invisible are able, under suitable conditions, to excite molecular movements corresponding to parts of the visible spectrum, and quite different in wave-lengths and in rapidity. This process is precisely the converse of what we perceive in thermal light; in the latter case the colors ascend, loaded with invisible heat rays; in the former they descend, loaded with invisible chemical rays, only noted, perhaps, by their actinic action on the photographic plate. Others, as the sulphide of calcium paints and the like, repeat their own vibrations for many hours, and we find in certain chemical salts of some rare metals, as lanthanum and cerium, the curious property of suddenly raising the whole scale, as in a recently introduced gas-lamp, in which a skeleton mantle of these oxides glows with a wondrously beautiful white light under the relatively low temperature of a small Bunsen burner; similar phenomena are manifested in the behavior of electric discharges in attenuated gases, as well as in what is known to children as "fox-fire," wood undergoing slow decomposition in damp places, or in the self-luminous secretions (corresponding, perhaps, to ptomaines or like products) of glow-worms and other animals. If we ever—as we probably soon shall—reach that point where we can illuminate our dwellings with "cold candles," as the inhabitants of tropical countries carry about a few fire-flies in a paper box for a lantern on dark nights, it must be by the study of these phenomena. But meantime "Old Sol" will continue to discharge his accumulating stores of both heat and light, for both these are essential, not only for use upon the planets, but throughout all the realms of space. In the transformation into and emission of his radiant energy the sun is not a chemical engine, but a mill,—one of those which " grind slowly, but they grind exceeding small."

The difference between radiated thermal light and heat is obviously one of degree only and not of kind. The undulations of light may be compared to the thrust of a rapier, and the more massive waves of radiant heat to the blow of a bludgeon, but the same resistance which arrests the advance of the one must retard and finally arrest that of the other, if sufficiently extended. Within the limits of a space in which Professor

Stewart conceives that the first rays of light which ever flashed forth at the dawn of creation, in the primal æons of the universe, are still to this day, along their original lines of radiation, "traversing space at the rate of one hundred and eighty-eight thousand miles per second," there must certainly be room enough and absorption enough (which even a few yards of mist will supply) to curb these runaway steeds somewhere along their lines of flaming passage. At that very point they are at work acting upon the molecules of the attenuated vapors of space, and assisting to re-establish the potential energy which has there been converted into another form of force by the planetary rotations of the solar systems of those distant regions. By the law of the diffusion of gases, and that of the diffusion or transference of heat-energy from molecule to molecule, the vast realms of interstellar space must tend to be all brought into approximate uniformity of tensions, and the force abstracted at those points of space occupied by the relatively few and insignificant solar systems will be returned, not directly at the identical places where such solar systems may exist, but at every part of space to which their radiant energy extends. As we give from our own supplies to other systems for their support, so they, in turn, give back again to us. It is said that in the earliest days of creation the stars sang together; they still sing together, planets and suns, as

> "Jura answers from her misty shroud
> Back to the joyous Alps, who call to her aloud."

When old Earth lifts his brimming beaker from the great crystal sea and drains it to the good health of all the stars of heaven, they each respond with fiery energy, and by their merry twinkle we may know how highly they appreciate the toast. We are all one family,—but what a family! Comets, planets, double stars, variable stars, stars of complementary colors, blue, yellow, orange, and red stars, stars which blaze up in sudden conflagration, apparently new stars, nebulæ half star and half vapor, nebulæ all vapor and others all stars, the vast milky-way like a wondrous river of hundreds of millions of solar systems, the insulated stars scattered through space like watchmen on the distant hills beyond the city walls, streams of stars, stars which are parting from each other in space like scattering families, and those which travel together in groups like pioneers in a strange country,—all these and doubtless other unknown types and forms compose this sidereal family. Will they fall into their categories as lawful subjects, so as to be properly classified in a single scheme of the visible order of creation, or shall we fail to interpret their apparent mysteries when we apply the same principles which have been successfully applied to the phenomena of our own solar system? Let us see.

In examining the sun, we find that a beam of its light passed through a prism is thrown upon the wall in a wedge-shaped streak of rainbow-tinted colors. Fraunhofer, many years ago, found that this spectrum was crossed at irregular intervals by a series of dark lines, of variable width and distance apart, of which he catalogued more than five hundred. These lines were subsequently found to correspond in the aggregate, in their position in the spectrum, with a series of bright lines of different colors which formed the separate spectra of various metals when burned, in vapor or powder, in the flame of an alcohol lamp. Each of these transverse lines was found to have a fixed and invariable position in the extended scale of the spectrum, and scarcely any lines of the different elements are alike; so that, when the spectrum is properly magnified under telescopic observation and the lines identified, we have the means of determining the presence or absence of such elements in the vaporous constitution of any incandescent body by examination of its spectrum. In this way many of our terrestrial elements are found to exist in the sun,—so many, in fact, that we know that the sun's nucleus, or core, must be composed substantially of the same elements, the same sort of matter, as exists on earth,—that we are, in fact, " a chip of the old block." But it was found—and this is the real basis of spectrum analysis—that if a certain metal or other element be burned in the flame of an alcohol lamp, and a more brilliant flame of the same metal or element burned in another lamp be observed through the first flame, it will be seen that, " while the general illumination of the spectrum is increased, the previous bright lines characterizing the element are now replaced by dark lines or lines relatively very faint; in a word, the spectrum characteristic of the given element is exactly reversed" (Appleton's Cyclopædia, article "Spectrum Analysis"). We have referred to this fact above in considering the origin of sun-spots, showing that they are due to increased heat acting upon the core of the sun so as to volatilize an abnormally large proportion of the elements usually in a more condensed state upon the surface of the solar body beneath its hydrogen envelope. These vapors, thus raised in temperature, are driven upward by their vola-

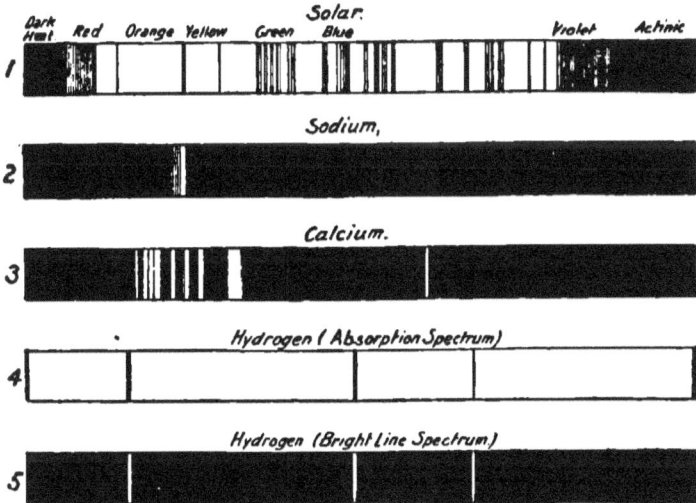

Spectra of different elements compared with the solar spectrum, and showing reversal of hydrogen lines under special circumstances.

tilization into the incandescent atmosphere of hydrogen, and the vaporous matters in the higher strata thus produce the characteristic absorption bands of these elements, while the overheated vapors, by a vast uprush from beneath, hurl aside the more highly heated hydrogen above to appear as faculæ around the sun-spot, the cooler upper layers of hydrogen following downward the subsiding vaporous metallic uprush as it sinks back beneath the photospheric level.

It is obvious that by similar spectrum analysis we may determine to a large extent the constitution of the fixed stars and other self-luminous bodies of space and interpret the phenomena which they exhibit. We quote the following from the previously cited article in Appleton's Cyclopædia, by Professor Proctor: "Spectroscopic analysis applied to the stars has shown that they resemble the sun in general constitution and condition. But characteristic differences exist, insomuch that the stars have been divided into four orders distinguished by their spectra. These are thus presented by Secchi, who examined more than five hundred star spectra: The first type is represented by Alpha Lyræ, Sirius, etc., and includes most of the stars shining with a white light, as Altair, Regulus, Rigel, the stars Beta, Gamma, Epsilon, Zeta, and Eta of Ursa Major, etc. These give a spectrum showing all the seven colors, and crossed usually by many lines, but *always by the four lines of hydrogen, very dark and strong.* The breadth of these four lines indicates a very deep, absorptive stratum at a high temperature and at great pressure. *Nearly half the stars* observed by Secchi [more than two hundred out of five hundred] showed this spectrum. The second type includes most of the yellow stars, as Capella, Pollux, Arcturus, Aldebaran, Alpha of Ursa Major, Procyon, etc. The Fraunhofer lines are well seen in the red and blue, but not so well in the yellow. *The resemblance of this spectrum to the sun* suggests that stars of this type resemble the sun closely in physical constitution and condition. About one-third of the stars observed by Secchi [more than one hundred and fifty out of five hundred] showed this spectrum. The third type includes Antares, Alpha of Orion, and Alpha of Hercules, Beta of Pegasus, Mira, and most of the stars shining with a red light. The spectra show bands of lines; according to Secchi, there are shaded bands, but a more powerful spectroscope shows multitudes of fine lines. The spectra resemble somewhat the *spectrum of a sun-spot*, and Secchi has advanced the theory that these stars are covered in great part by spots like those of the sun. About one hundred [out of five hundred] of the observed stars belong to this type." (It should be noted that the presence of sun-spots is no evidence of diminished heat in a sun; see Professor Proctor in his "Myths and Marvels of Astronomy," article "Suns in Flames:" "It may be noticed, in passing, that it is by no means certain that the time when the sun is most spotted is the time when he gives out least light. . . . All the evidence we have tends to show that when the sun is most spotted his energies are most active. It is then that the colored flames leap to their greatest height and show their greatest brilliancy, then also that they show the most rapid and remarkable changes of shape.") . . . "The fourth type differs from the preceding in the arrangement and appearance of the bands. It includes only faint stars. A few stars, as Gamma of Cassiopeia, Eta of Argus, Beta of Lyra, etc., show the *lines of hydrogen bright instead of dark*, as though surrounded by hydrogen glowing with a heat more intense than that of the central orb itself around which the hydrogen exists."

All the above five hundred stars reveal the presence of hydrogen under precisely such conditions as conform to the general principle involved in the source and mode of solar energy as herein stated. But a single star (Betelgeuse) was observed by Huggins and Miller in England which showed the lines of sodium, magnesium, iron, bismuth, and calcium, "but found those of hydrogen wanting." Of the spectrum of this gas, Professor Ball says, "The hydrogen spectrum appears to present a simplicity not found in the spectrum of any other gas, and therefore it is with great interest that we examine the spectra of the white stars, in which *the dark lines of hydrogen* are unusually strong and broad." Referring to the new star in the Northern Crown, which burst forth in 1866, the same writer says, "The feature which made the spectrum of the new star essentially distinct from that of any other star that had been previously observed was the presence of *certain bright lines* superposed on a spectrum with dark lines of one of the ordinary types. The position of certain of *these lines showed that one of the luminous gases must be hydrogen.*" Of this particular star (Betelgeuse) it is said (Proctor's "Familiar Essays"), "Red stars and variable stars affect the neighborhood of the Milky Way or of well-marked star-streams. The constellation Orion is singularly rich in objects of this class. It is here that the strange 'variable' Betelgeuse lies. At present this star shows no sign of variation, but a few years ago it exhibited remarkable changes." We thus see that Betelgeuse is a variable star, and it must have passed in its different variations between the limits of extreme brilliancy, in which the lines of hydrogen appear bright, and that of a less brilliant stage, in which they appear dark,—that is, as absorption bands. It has thus, in fact, run the gamut, so to speak, of color changes, and now occupies an intermediate position in the scale. In his article " Star unto Star," the same writer says, " On this view we may fairly assume that the darkness of the hydrogen lines is a characteristic of stars at a much higher temperature than our sun and suns of the same class." We have already seen that the spectra of stars of the fourth type—Appleton's Cyclopædia, " Spectrum Analysis "—" show the lines of hydrogen bright instead of dark, as though surrounded by hydrogen glowing with a heat more intense than that of the central orb itself." Professor Dunkin says, in his work " The Midnight Sky," " One of the conclusions drawn by Kirchhoff from these experiments is that each incandescent gas *weakens*, by absorption, rays of the same degree of refrangibility as those it emits; or, in other words, that the spectrum of each incandescent gas

160 SOURCE AND MODE OF SOLAR ENERGY.

is reversed when this gas is traversed by rays of the same refrangibility emanating from an intensely luminous source which gives of itself a continuous spectrum like that of the sun." . . . "The third division, including Betelgeuse, Antares, Alpha Herculis, and others of like color, seems to be affected by something peculiar in their physical composition, *as if their photospheres contained a quantity of gas at a lower temperature than usual.* The stars in this class have generally a ruddy tint, probably owing to their light having undergone some modification while passing through an absorbing atmosphere. . . . A great number of the stars in the third division are variable in their lustre." We

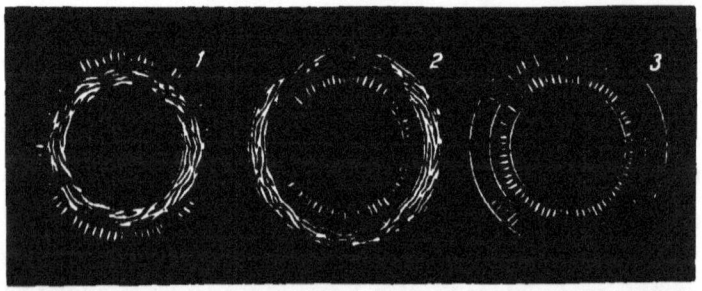

Reversal and neutralization of spectroscopic lines in spectrum of a variable star like Betelgeuse.—1, photosphere hotter than chromosphere; hydrogen lines dark. 2, chromosphere hotter than photosphere; hydrogen lines bright. 3, chromosphere and photosphere equally incandescent.

may therefore readily conclude that midway between the inverted lines which constitute the dark absorption bands and the faint spectra which show the bright lines of hydrogen direct there must be an atmosphere of glowing hydrogen superposed upon a deeper one in such proportion that it will

not reveal its presence in the spectroscope at all;
for when the dark and light bands, which occupy
precisely the same position in the spectrum, are of
approximately equal intensity the result will obviously be the neutralization of both. That among
a myriad suns, some with dark hydrogen lines and
some with bright, there should occur occasionally
an example corresponding to this point of divergence, and especially among variable stars, is not
only to be expected, but is, in fact, confirmatory of
the general hypothesis itself. It is an exception
which emphatically proves the rule, when we can
trace the operative cause which has produced it.

CHAPTER VI.

THE PHENOMENA OF THE STARS.

LET us now consider the phenomena of the double stars. These were formerly believed to be single orbs, but the more powerful telescopes of recent years have shown them to consist of two suns, each substantially similar to our own sun, revolving around each other at a relatively small distance apart. In Appleton's Cyclopædia, article "Star," we read, "It is noteworthy that few simple stars show such colors as blue, green, violet, or indigo; but among double and multiple star systems not only are these colors recognized, but such colors as lilac, olive, gray, russet, and so on. A beautiful feature in many double stars remains to be noticed: it is often found that the components exhibit complementary colors. *This is oftener seen among unequal doubles*, and then the larger component shows a color from the red end of the spectrum, as red, orange, or yellow, while the smaller shows the corresponding color from the blue end, as green, blue, or purple. The colors are real, not merely the result of contrast, for when the larger star is concealed the color of the smaller remains (in most cases) unchanged. Spectrum analysis shows that the colors of many double stars are due to the absorptive vapors cutting off certain portions of the light. . . . The components are circling around each other, or rather around their common center of gravity." Professor Ball, in his work "In the High Heavens," says, "There is no more pleasing phenomenon in sidereal astronomy than that presented by the contrasted hues often exhibited by double stars. . . . It seemed not at all impossible that there might be some optical explanation of colors so vividly contrasted emanating from points so contiguous. It was also remembered that blue stars were generally only present as one member of an associated pair. . . . When, however, Dr. Huggins showed that the actual spectrum of the object demonstrated that the cause of the color in each star arose from absorption by its peculiar atmosphere, it became impossible to doubt the reality of the phenomena. Since then it has been for physicists to explain why two closely neighboring stars should differ so widely *in their atmospheric constituents*, for it can be no longer contended that their beautiful hues arise from an optical illusion."

Of these double stars with complementary colors we quote the following from Professor Dunkin (who, in turn, quotes from Admiral Smyth, the author of "Sidereal Chromatics"): "In Eta Cassiopeiæ the large star is a dull white and the smaller one lilac; in Gamma Andromedæ, a deep yellow and sea-green; in Iota Cancri, a dusky orange and a sapphire blue; in Delta Corvi, a bright yellow and purple; and in Albiero, or Beta Cygni, yellow and blue. In most of the remaining stars of the list the contrasting colors are equally marked, and also in many others which are not included in it." Some of these double stars are variable in their colors, as are the ordinary single variables, and, of course, for a similar reason,—to wit, the varying intensity of more or less cumulative planetary impacts.

The interpretation, of course, as explained below, is that these suns, each one of different mass and consequently of different electrical resistance, are arranged in parallel circuit along a single line of electric current; a pair of different-sized arc or incandescent lamps, similarly arranged, would exhibit precisely the same phenomena. A compound solar system of this sort, apparently, with double sun and single planetary system in process of formation, nearly completed from a spiral nebula, is shown in a gaseous nebula within the constellation Ursa Minor, illustrated in Lord Rosse's drawing (see Nichol's "Architecture of the Heavens," Plate X., lower figure).

Reduced from Plate X. of Nichol's work. For interpretation see Chapter XIII., "The Genesis of Solar Systems."

More than three thousand of these binary stars have been catalogued, and some of them make a complete revolution about their common centers of gravity—so distant are they from each other—in periods of not less than sixty, or even eighty, years.

Of the double star Mizar,—the middle one of the three which form the tail of the Great Bear,—Professor Ball states that, by new methods of spectroscopic analysis, the component stars which form this double have been found to be one hundred and fifty millions of miles apart, while Alcor, a smaller star, visible to the naked eye, and enormously farther from Mizar than are the components of the latter from each other, moves through space in a parallel direction and with the same velocity as its double companion. What the connection may be, if any, we do not know, but their identical course is obviously related to some common circumstance of origin, as is the probable case with those other groups of stars which drift through space together. They show that solar systems are not necessarily individual creations, but may be formed in groups at the same period of time, and by the operation of natural laws simultaneously directed upon or into the creative matter from which solar systems are built up and sent along their way. It has been already shown that our sun has a motion around the center of gravity of our own solar system, as a whole, similar to that of the binary stars around each other, but that, by reason of his vast relative mass (seven hundred and fifty to one for all the planets), this center is always within the confines of his own volume. If, however, our sun were divided into two suns one, two, or five million miles apart, each revolving around a common center of gravity situated between the two, and the planets revolving around the same center of gravity, but relatively more distant, the planets would thus rotate around both suns as a common center, and with the electric polarity of both suns the same, as must necessarily be the case, they would present phenomena precisely similar to those exhibited by the double stars. And such might very easily be the case in even a system so small as our own, for the planet Mercury has so elliptical an orbit that its distance from the sun varies in different parts of its annual movement from twenty-eight to forty-five millions of miles. There would then be mutual electric repulsion of the two solar electrospheres, such as we see in the case of comets and in the sun's corona and long streamers. Professor Proctor, article "The Sun's Long Streamers," says, "These singular appendages, like the streamers seen by Professor Abbe, extend directly from the sun, as if he exerted some repellent action. . . . I cannot but think that the true explanation of these streamers, whatever it may be (I am not in the least prepared to say what it is), will be found whensoever astronomers have found an explanation of comets' tails. . . . Whether the repulsive force is electrical, magnetic, or otherwise, does not at present concern us, or rather does concern us, but at present we are quite unable to answer the question." A similar example is to be found in the self-repellent positive electrospheres of the earth and moon, illustrated on a previous page, which, in fact, are types among planets of precisely what we find in double stars. Now, if these double central suns, with a common system of planets revolving around them both, differ one from the other in size, they will differ also in the depth and density of their hydrogen atmospheres, and the electric forces directed

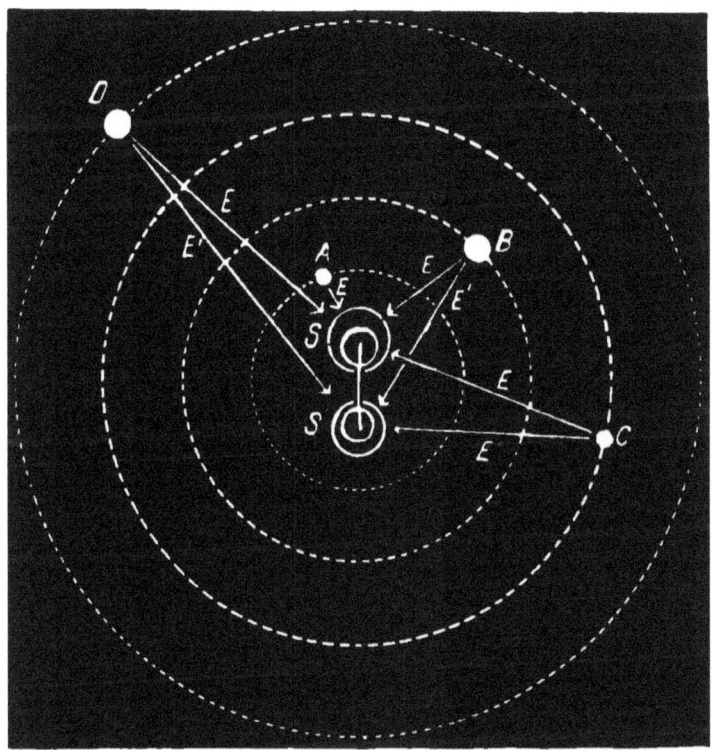

Double stars with complementary colors.—A, B, C, D, planets; S, S', double central sun; S, larger sun, with dark absorption spectrum, yellow-red, or orange; S', smaller sun, many bright lines, bluish-white; E, E', lines of planetary energy; S, S' also show self-repulsion of their solar electrospheres.

against them will produce different results in each. In one we will have high temperature, great volatilization, and wide absorption bands; in the other, a shallow atmosphere, a temperature below that of an extensive volatilization of its metallic components, and a spectrum rich in light at the blue end, while the former one will be correspondingly richer in the yellow and red rays at the opposite and lower end of the spectrum. One, in fact, will manifest the phenomena of blue-white stars, the other, those of orange-red, but variously modified in a chromatic series. The case may be extended to multiple stars, and complementary colors, more or less perfect, may be almost predicated as the law of compound solar bodies having cores like that of our sun, but each of different mass, and surrounded by hydrogen atmospheres of different depths and densities, both acted upon by the same exterior planetary electrical currents. It is certainly true of double stars, and probably so of all the others. Of course such enormously massive double suns presuppose enormous planets, rotating around them at enormous distances; but when we compare the distance of our own satellite, the moon, from the earth with the distance of Neptune from the sun, and consider that the light of the sun will reach Neptune in about four hours, and then compare this distance with the inconceivable distances of space requisite to retard and merge all radiant energy into the diffused molecular energy of position, our wonder will cease.

We have also to consider those single stars which (see Appleton's Cyclopædia, article "Star") are variable in their brilliancy. "These stars may be divided into periodic variables, irregular variables, and temporary stars. Periodic variable stars are those which undergo increase and diminution of light at regular intervals. Thus, the star Mira, or Omicron of Cetus, varies in lustre, in a period of three hundred and thirty-one and one-third days, from the second magnitude to a faintness such that the star can only be seen with a powerful telescope, and thence to the second magnitude again. It shines for about a fortnight as a star of the second magnitude, and then remains invisible for five months, the *decrease* of lustre occupying about three months, the *increase* about seven weeks. Such is the general course of its phases. It does not always, however, return to the same degree of brightness, nor increase and diminish by the same gradations; neither are the successive intervals of its maxima equal. From recent observations and inquiries into its history, the mean period would appear to be subject to a cyclical fluctuation embracing eighty-eight such periods, and having the effect of gradually lengthening and shortening alternately those intervals to the extent of twenty-five days one way and the other. The irregularities in the degree of brightness attained at the maximum are probably also periodical. . . . It suggests a probable explanation of these changes of brightness, that when the star is near its minimum, its color changes from white to a full red, which, from what we know of the spectra of colored stars, seems to indicate that the loss of brightness is due to the formation of many spots over the surface of this distant sun.

"Algol is another remarkable variable, passing,

however, much more rapidly through all its changes. It is ordinarily a second-magnitude star, but during about seven hours in each period of sixty-nine hours its lustre first diminishes until the star is reduced to a fourth magnitude, and after it has remained twenty minutes at its minimum its lustre is gradually restored. It remains a second-magnitude star for about sixty-two hours in each period of sixty-nine hours. These changes seem to correspond to what might be expected if a large opaque orb is circling around this distant sun in a period of sixty-nine hours, transiting its disk at regular intervals."

Of this star, Professor Ball says, " Applying the improved spectroscopic process to Algol, he [Vogel] determined on one night that Algol was retreating from the earth at a speed of twenty-six miles per second. . . . When Vogel came to repeat his observations, he found that Algol was again moving with the same velocity, but this time towards the earth instead of from it. . . . It appeared that the movements were strictly periodic; that is to say, for one day and ten hours the star is moving towards us, and then for a like time it moves from us, the maximum speed being . . . twenty-six miles a second. . . . It is invariably found that every time the movement of retreat is concluded the star loses its brilliance, and regains it again at the commencement of the return movement. . . . The spectroscopic evidence admits of no other interpretation save that there must be another mighty body in the immediate vicinity of Algol. . . . Algol must be attended by a companion star which, if not absolutely as devoid of intrinsic light as the earth or the moon, is nevertheless dark relatively to Algol. Once in each period of revolution this obscure body intrudes itself between the earth and Algol, cutting off a portion of the direct light from the star and thus producing the well-known effect." This is, in fact, a periodic transit or eclipse of Algol by a planet, such as we see in eclipses of our own sun by the moon and the inner planets, except that Algol's planet is apparently single like our moon with reference to the earth, and that it is relatively much larger than any of our own planets, as we would necessarily suppose it to be, if solitary. Its mass has been computed by the effects which it produces, and we learn that it is not a dark sun with a brilliant planet, but a brilliant sun with a dark planet, just as our solar system presents. " Algol, at the moment of its greatest eclipse, has lost about three-fifths of its light; it therefore follows that the dark satellite must have covered three-fifths of the bright surface. . . . The period of maximum obscuration is about twenty minutes, and we know the velocity of the bright star, which, along with the period of revolution, gives the magnitude of the orbit." From these data it has been computed that the globe of Algol itself is about one-fourth larger than that of our visible sun, but its mass is so much less that its weight is only one-half that of our sun, so that its body is probably gaseous. The author concludes, "No one, however, will be likely to doubt that it is the law of gravitation, pure and simple, which prevails in the celestial spaces, and consequently we are able to make use of it to explain the circumstances attending the movements of Algol's dark companion. *This body is the smaller of the two*, and the speed with which it moves is double as great as that of Algol, so that it travels over as many miles in a second as an express train can get over in an hour. The companion of Algol is about the same size as our sun, but has a mass only one-fourth as great. This indicates a globe of matter which must be *largely in the gaseous state*, but which, *nevertheless, seems to be devoid of intrinsic luminosity*. Their distance [apart] is always some three million miles. This is, however, an unusually short distance when compared with the dimensions of the two globes themselves." With this exception, the author says, "the movements of Algol and its companion are not very dissimilar to movements in the solar system with which we are already familiar." It will be seen that the want of luminosity in the dark companion of Algol finds a ready explanation in the fact that it is a planet, acting precisely as our own planets do, and that the luminosity of Algol itself is directly attributable to the electricity developed by the presence of this planet rotating axially and orbitally around it, and the darkness of the planet itself is the necessary correlative of the heat and light of its sun. The planet has about one-half the density of Saturn, while Algol has one-half the density of the sun, and hence we should expect to find on Algol an atmosphere largely composed of glowing hydrogen, and on its planet an atmosphere largely composed of oxygen, in which, doubtless, float enormous clouds of aqueous vapor. The interpretation is direct and conclusive, and upon no other hypothesis can the facts be explained, for their close connection with each other demonstrates their common origin, and their masses are not so different one from the other as to permit, on any theory of their coequal origin as suns, one to glow with the fires of youth and energy and the other to have grown dark and dead from old age and exhaustion, and especially so if still in its gaseous stage, which is that which must characterize its highest state of incandescent energy from the most active condensation of its volume, if the nebular hypothesis has any validity whatever. In fact, this example alone, if the constitution of Algol's dark satellite is really gaseous, must go very far to throw the gravest doubt, in itself, on the validity of this hypothesis.

The star Beta, of the constellation Lyra, has a full period of twelve days and twenty-two hours, divided into two periods of six days and eleven hours, in each of which the star has a maximum brightness of about the three and one-half magnitude, but in one period the minimum is about the four and one-third magnitude, while in the other it is about the four and one-half magnitude. This peculiarity points, it is said, to an opaque orb with a satellite, the satellite being occulted by the primary in the alternative transits, and therefore the loss of light is less.

The star Delta of Cepheus is quite different, however, for, while it takes only one day and fourteen hours in passing from its minimum to maximum of brightness, it occupies three days and nineteen hours, or somewhat more than double this time, in passing from maximum to minimum. Two or three hundred of these variable stars are already known. The above examples are cited in detail because they furnish the strongest possible proof of the truth of the hypothesis which we are endeavoring to present. While the movements of the stars Algol and Beta Lyræ may find an adequate interpretation in the one case in a large occulting planet, and in the other in an occulting planet with a satellite, it is obvious that Mira and Delta Cephei cannot be explained except by the presence of planetary bodies or satellites which do not *mechanically* occult the light of their suns. In these regularly variable stars it is the light which varies, but of course the solar heat must vary also,—that is to say, the solar energy varies regularly, but with unequal periods of growth and decline and with larger periods of cyclical variation in addition. Such variations can only be produced by the action of permanently connected and orbitally rotating planetary bodies, acting *dynamically* through space, to regularly increase and diminish the solar energy, and such bodies can only do this by their orbital positions with reference to each other and to the central sun itself. In this case, since the activity of solar energy is most unquestionably varied by the planetary energies,

THE PHENOMENA OF THE STARS. 175

by their position and movements, at least a portion of solar energy *must* be due to planetary action, and if this be so, it may be affirmed with certainty that substantially all solar energy may be produced in the same way; for, otherwise, we seek for two diverse causes to produce a single effect, which may be produced by one. We have no knowledge, however, of any planetary energy which could operate to increase or diminish the energy of the central sun in its emission of light, except that which we have already presented, and no theory of our own sun's energy hitherto advanced has ever taken cognizance of the planetary energies of our system as an effective cause for those of the sun. But while the sun's energy is—as it must be in this case—the outcome of that of the planets, it is equally obvious that the planets themselves can have no permanent, inherent energy of their own to generate or modify such energy of the sun, since they are in fact supplied by the solar energy, and their motions are controlled and regulated by the sun itself. Hence the inference is irresistible that the planets must derive their primary force from an external source not solar, and this they can only do by means of their rotation in space, and the only force derivable from space of which we have any knowledge is electricity, so that the circle thus becomes complete. How now shall we explain these periodical aberrations of energy? The color of a star, as we know, is no criterion of its age or size. The color is due to atmospheric absorption of the radiant light. The double stars, for example, revolve around each other at regular periods, and they are necessarily of nearly the same age, as sidereal ages are computed, but they frequently differ one from the other in color, and multiple stars may be all different each from the others; and the color, as before stated, is no criterion of size, for a small sun, with its glowing hydrogen in a state of high incandescence, and with few absorption bands in its spectrum, will appear bluish-white, or of that specific type of stars, without reference to size, while a much larger sun, with its light darkened by broad absorption bands and sun-spots, will appear orange or red; and, consequently, difference of color can be no criterion of distance, since a blue-white star of small size will outshine a red orb of much greater magnitude, whether it be more or less distant. The variable stars, for these reasons, belong to the order of red stars mostly, if not altogether. We must also bear in mind that sun-spots do not diminish the solar heat, as they are the result of increased and not of diminished energy. Electric currents of high potential pass directly, as we know, along the lines of least resistance to their opposite center of polarity, so that two planets nearly in conjunction with each other transmit their currents almost directly towards the sun's center, and upon the same point of solar latitude, while, if at right angles with the sun, they must deliver their electricity along converging lines and thus strike the solar surface at different points. Currents of electricity of high potential also (see "Electricity in the Service of Man," page 75), by their own passage, facilitate the passage of succeeding currents, so that generators discharging along the same lines find less and less resistance. It is true that we find no appreciable resistance in the passage of these currents between the earth and the sun, as their velocity is that of light, but both light and electricity may be equally retarded by resistance in a small degree. We know also that in the condensed hydrogen atmosphere of the sun there must be resistance, and also that the resistance in fluids diminishes as the temperature rises. Considering now the variable star Mira, as above described, we observe, as is the case with Delta Cephei, also cited, that the period between its greatest light, in a descending scale, and its least is about twice as long as its rise from minimum to maximum. During a period of four years (1672 to 1676) it is said that it was not visible at all.

If Mira be considered a relatively small sun, with its axis strongly inclined to the planetary plane, and having three planets only, two of them constituting a double planet, like the earth and moon, but nearly equal in size, and having a rotation about the sun in nearly eleven months and a rotation about each other in the same period, and, besides these, a much more distant large planet, something like our Jupiter, with an orbital period of many years, so that the cycle of relative positions is complete in about eighty-eight of the shorter periods of variation, we would have such results as we see in Mira. Twice in each revolution of the double planet its two members and their sun would be in conjunction, and we would have great brilliancy and whiteness until the metallic elements began to volatilize in increased pro-

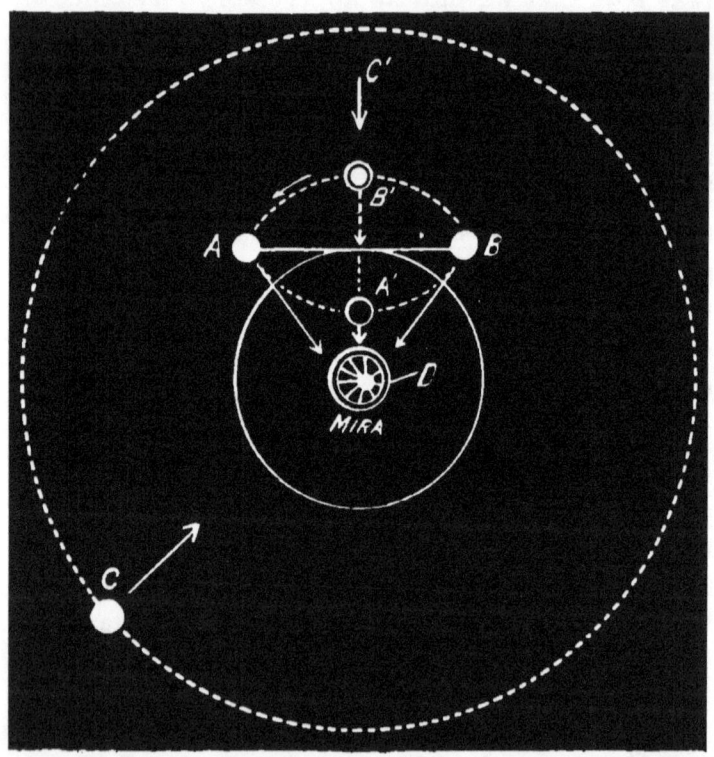

Possible solar system of variable star Mira.—D, central sun with axis of rotation considerably inclined from perpendicular to planetary plane; A, B, double internal planet, like the earth and moon, with short orbital period; C, large external planet, like Jupiter, with long period; line A', B', C', conjunction, period of greatest energy; A, B, C, opposition, period of least planetary energy.

portions; then an era of wide absorption bands and redness, gradually increasing to a maximum after its periods of greatest light, and then slowly diminishing as the double planet advanced in its

THE PHENOMENA OF THE STARS. 179

rotation; and, finally, as it again approached conjunction, the brilliant hydrogen illumination, subsequently followed by the gradually darkened spectrum, and so on, while the large outer planet by its various positions would first relatively retard and then accelerate the variation until its grand cycle was complete. The permanent disappearance for years, if true, may be due to other causes, which will be referred to in considering the phenomena of new and temporary stars. Many of the irregular variables may doubtless be similarly explained,—our own sun, in fact, being a variable with a period of about eleven years,—and doubtless the apparent irregularity in most cases is due to lack of sufficient time for observation. Those stars which are in fact really irregular in their variation owe their changes, doubtless, to the same causes which produce new stars, so called, and "suns in flames," which will be next considered.

Among the countless stars of heaven a great catastrophe seems occasionally to occur. A star bursts out into sudden flame, to all appearance, or a great fixed star appears where no star had ever been seen before. In Professor Proctor's article, "Suns in Flames" ("Myths and Marvels of Astronomy"), we will find an extended discussion of these wonderful phenomena. The astronomer Tycho Brahe described the one which appeared in 1572 as follows: "It suddenly shone forth in the constellation Cassiopeia with a splendor exceeding that of stars of the first magnitude, or even Jupiter or Venus at their brightest, and could be seen by the naked eye on the meridian at full day. Its brilliancy gradually diminished from the time of its first appearance, and at the end of sixteen months it entirely disappeared, and has never been seen since. During the whole time of its apparition its place in the heavens remained unaltered, and it had no annual parallax, so that its distance was of the same order as that of the fixed stars." Tycho described its changes of color as follows: first, as having been of a bright white; afterwards of a reddish-yellow, like Mars or Aldebaran; and, lastly, of a leaden white, like Saturn. In 1604 a first-magnitude star suddenly appeared in the right foot of Ophiucus. "It presented appearances resembling those shown by the former, and disappeared after a few months." Many other cases are cited by astronomers, and in 1866 "a star appeared in the Northern Crown, the observations of which threw great light on the subject of so-called new stars. In the first place, it was found that where this new star appeared there had been a tenth-magnitude star; the new star, then, was in reality a *star long known, which had acquired new brilliancy.* When first observed with this abnormal lustre, it was shining as a star of the second magnitude. Examined with the spectroscope, its light revealed a startling state of things in those remote depths of space. The usual stellar spectrum, rainbow-tinted and crossed by dark lines, *was seen to be crossed also by four exceedingly bright lines, the spectrum of glowing hydrogen.* . . . The greater part of the star's light manifestly came from this glowing hydrogen, though it can scarcely be doubted that the rest of the spectrum was brighter than before the outburst, the materials of the star being raised to an intense heat. The maximum brightness exceeded that of a tenth-magnitude star nearly eight hundred times. After shining for a short time as a second-magnitude star, it diminished rapidly in lustre, and it is now between the ninth and tenth magnitudes" (Appleton's Cyclopædia). Of this new star, Professor Ball says, "Another memorable achievement in the early part of Dr. Huggins's career is connected with the celebrated new star that burst forth in the Crown in 1866. It seemed a fortunate coincidence that just at the moment when the spectroscope was beginning to be applied to the sidereal heavens a star of such marvellous character should have presented itself. . . . The feature which made the spectrum of the new star essentially distinct from that of any other star that had been previously observed was the presence of certain bright lines superposed on a spectrum with dark lines of one of the ordinary types. The position of certain of these lines showed that one of the luminous gases must be hydrogen. . . . The spectroscope showed that there must have been something which we may describe as a conflagration of hydrogen on a stupendous scale, and this outburst would account for the sudden increase in luminosity of the star, and also to some extent explain how so stupendous an illumination, once kindled, could dwindle away in so short a time as a few days." It will be seen that these new stars

182 SOURCE AND MODE OF SOLAR ENERGY.

leap suddenly into great brilliancy: it is a matter of a few hours only. After remaining a very short time in this stage of abnormal incandescence, they gradually die out again in lustre and revert *to their original condition;* they are not consumed either in body or atmosphere.

Several theories have been advanced to account for these remarkable phenomena; see "Suns in Flames," by Professor Proctor. One is, in effect, that by some sudden "internal convulsion a large volume of hydrogen and other gases was evolved from it, the hydrogen by its combination with some other element giving out the lines represented by the bright lines, and at the same time heating to a point of vivid incandescence the solid matter of the star's surface. . . . As the liberated hydrogen gas became exhausted the flame gradually abated, and with the consequent cooling the star's surface became less vivid and the star *returned to its original condition;*" which, by the way, it never could have done if its atmosphere had been exposed to such a disintegration, without the construction of an entirely new atmosphere precisely similar to the one just destroyed. The process would be one of simple combustion. It requires the evolution of enormous volumes of hydrogen from within the planet, and of other enormous volumes of something else, by which to burn it up and yet not burn up the *original* hydrogen envelope. This other element could not have previously existed outside the solar body and contiguous thereto, or it would have burned up the ordinary hydrogen envelope of the sun long before, as well as the metallic vapors floating therein. Both these mutually hostile gases must have come from within, and this is manifestly impossible, as we should thus have explosion and solar destruction, but not combustion. There is no reason to believe that hydrogen, the lightest of elements, could have remained occluded within the solar mass, to the exclusion of the heavier metals, if disassociated, and if held combined no such sudden liberation could occur. Besides, such convulsion would be impossible in any sun at all resembling ours, as any further liberation of gases from internal condensation must be due to solar contraction, hence gradual, and not sudden. Moreover, such liberation of hydrogen gas from within would show its spectrum loaded, at its earliest eruption, with absorption bands; and, finally, the convulsion presupposes as great an activity, and consequently as great a difficulty, before the phenomenon as the phenomenon itself presents; for such vast disturbance of mass would be more difficult to account for, and require more energy to produce, than the results themselves. Moreover, the whole mass of the star appeared to increase equally in temperature, as shown by the spectrum, and, if produced by an internal convulsion, this must have extended to, if not proceeded from, its core; so that while the combustion of hydrogen might have ceased in a very brief time, the intense heat of the solar mass could not have been dissipated for thousands of years. It would, in fact, have disrupted the whole orb.

Another theory is that this vast incandescence was caused by the "violent precipitation of some mighty mass—perhaps a planet—upon the globe of that remote sun, by which the momentum of the falling mass would be changed into molecular motion; in other words, into heat and light." This theory is no more plausible than the other, since it fails to account for the enormous volume of hydrogen, with bright lines, as a result of such contact; while Professor Proctor very clearly shows that such contact would have been preceded, necessarily, by repeated partial grazings, as the outside body repeatedly passed in swifter and closer passage by the sun in its gradually approaching orbital revolutions, and that the increase of light and heat must have been measured by years instead of by hours. The same difficulties exist in the supposed passage of the star through nebulæ or star clouds, of which Professor Proctor says, " As for the rush of a star through a nebulous mass, that is a theory which would scarcely be entertained by any one acquainted with the enormous distances separating them. . . . All we certainly know suggests that the distances separating them from each other are comparable with those which separate star from star." In fact, no tenable theory has been advanced which will cover the phenomena. Professor Proctor describes a star which flamed out in 1876. At midnight, November 24, a star of the third magnitude was noticed in the constellation of the Swan; its light was very yellow; its brilliancy rapidly faded. On December 2 it was equal to a star of the fifth magnitude only, and the color, which had been yellow, was now greenish-blue. "The star's spectrum at this time consisted almost entirely of bright lines. December 5 he found three bright lines of hydrogen, the strong double line of sodium, the triple line of magnesium, and two other lines. One of these last seemed to agree exactly in position with a bright line belonging to the corona seen around the sun during total eclipse." The star afterwards faded away gradually until quite invisible to the naked eye. It will be noticed that none of the above elements—sodium, potassium, or magnesium—are such as would combine with hydrogen to produce the phenomena in question. Professor Proctor concludes, "This evidence seems to me to suggest that the intense heat which suddenly affected this star had its origin from without." He suggests possible meteoric flights; but meteoric stones themselves are separated in space by enormous distances, and these, if converged in orbital flight, would present the same phenomena of successive grazings as a small planet approaching under like circumstances, and by their gradually increasing incandescence we should certainly have other elements visible in the spectroscope besides those observed. And these meteoric bodies, if projected into the sun, would pass in a very brief time through the hydrogen envelope, producing only local phenomena, so that their first blow would be manifested in volatilization of the outer portions of the mass and broad absorption bands, and consequent redness of the planet, exhibiting great heat, but not great light. In such case the bright lines of hydrogen, if they appeared at all, would only be visible as an after-consequence, and not at the earliest moment of conflagration,—that is, the star might grow from red to white, but by no possibility the reverse. It is, however, characteristic of these new stars that their first flash, as it were, is into the incandescence of directly glowing hydrogen, with its bright lines, then through a series of gradually increasing sun-spots, and finally a slow return to their original condition and apparent magnitude. It is obviously a surface phenomenon of the solar atmosphere, primarily, then followed by consequences involving only the outer surface of the solar core, but with no observable permanent change in the character or constitution of the mass of the sun itself. These characteristics are invariable, and the sequence of phenomena is the same in all the cases observed.

CHAPTER VII.

TEMPORARY STARS, METEORS, AND COMETS.

WHAT, then, is the probable cause of these terrific conflagrations, as they appear to us? Take an ordinary electric induction machine,—a Holtz or a Wimshurst,—and, if the surrounding air is moist, as we operate it we will find that the results are poor, the sparks short and relatively few; but let us take the machine into another room in which the atmosphere is dry and crisp. A wondrous change will occur, and instead of a current which could scarcely flash across a few inches of space, we will now have so great an increase of energy that its tension will even cause the spark to perforate and destroy the glass walls of the heavy Leyden jars in which it is condensed. The vast realms of space, with their attenuated vapors, are the field in which the planetary electric generators operate, and into which, likewise, myriads of suns constantly pour their light and heat. We may consider this space, according to the popular view, to be uniform in constitution and density throughout all its parts,—that it is, in fact, like a vast, silent, and motionless dead sea. But this cannot possibly be true, any more than throughout the vast compass of our own atmosphere; for while some parts of space are peopled by millions of solar systems, others, as we can plainly see, so far as telescopic vision extends, are comparatively vacant. Far more electricity is being abstracted (so to speak) in some parts of space than in others, and far more heat and light are being poured back to restore the equilibrium in some than in others. We have already seen that the temperature at the exterior surface of the terrestrial atmosphere is estimated to be more than two hundred degrees higher than in the realms of open interplanetary space; hence there must be currents,—currents of rotation like cyclones, vortical currents like whirlwinds, currents of transmission like our land- and sea-breezes and the trade-winds,—and, in fact, all space must be in a state of constant displacement and replacement, and, if visible, we should see it like a vast room filled with smoke, in which currents of every shape and direction and of all velocities would be manifest. Such currents could throw nebulæ during their condensation into rotation which could never rotate of their own motion, or gather to centers of aggregation vast whirling clouds of spatial matter, and in the spiral nebulæ we may see many such movements of rotation in apparent active progress. Of these we read in Appleton's Cyclopædia, "They have the appearance of a maelstrom of stellar matter, and are among the most interesting objects in the heavens." In Professor Nichol's splendid work ("The Architecture of the Heavens," 1850) we may see magnificent engravings of these wonderful phenomena, from the drawings by Lord Rosse,

TEMPORARY STARS, METEORS, AND COMETS. 189

and no one can study these figures without realizing the presence of vast currents in space.

In the great spiral nebula in the constellation *Canes Venatici* (see illustration in Chapter XII.) we perceive that the tail of the smaller nebula has been drawn into the outer convolution of the great spiral, against the radial repulsion of the latter 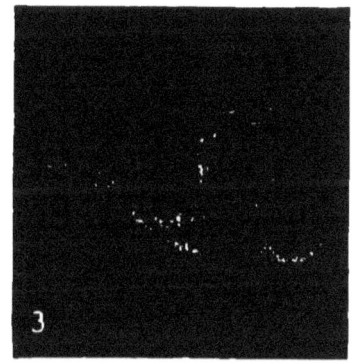 nebula, as we can see by its curvature. This can only be due to a tremendous inflowing current in space. Were the deflection due to gravity the trend would be to the center and not to the outer convolution of the larger nebula. Professor Nichol says, "The spiral figure is characteristic of an extensive class of galaxies." Not only in the spiral, but in other forms of nebulæ we may observe these currents of space, so that we cannot fail to perceive that they exist, and we should even conclude, *a priori*, that these must exist.

In the elongated linear nebula in Sobieski's Crown, illustrated above, its length is deflected into irregular curves apparently due to counter-currents of space. These gaseous nebulæ, Flammarion says, "appear like immense vaporous clouds tossed about by some rough winds, pierced with deep rents, and broken in jagged portions." It may be said generally that every sun, as it drifts through space, must leave a wake of increased electric potential among the molecules which line its pathway. Beyond the limits of every vortex extend radial or tangential, polar or equatorial, streams of space, and these must extend without limit until deflected or neutralized by other conditions. Throughout all space, just as in our own atmosphere, but vastly more slowly, there must be an infinitude of movements in every direction,—movements in lines, circles, vortices, ellipses and irregular curvatures, and of all possible varieties of mass and volume.

Suppose, now, a sailing vessel lighted with incandescent lamps, the electrical currents for the support of which are derived from the chemical action of sea-water on multiple pairs of suitable metallic plates arranged to extend downward as a galvanic battery into the ocean as the ship sails along, and that these plates, by the chemical action of the sea-water at ordinary temperatures, should furnish a sufficient current to properly light the vessel. If the constancy of such current depended on the average temperature of the sea-water, at, say, sixty degrees Fahrenheit, we should find that, on suddenly crossing into the Gulf Stream, with a temperature twenty degrees higher, the energy of the battery would be rapidly increased and the lights would glow with increased brilliancy until, on emerging from the Gulf Stream at its opposite side, the original status would be gradually restored. If these distant solar systems, in their drift through space, should encounter a corresponding stream under an increased molecular tension, more highly heated, for example, or charged with electrical potential by the surrounding solar systems, or otherwise, we should expect a similar result to ensue,— that the currents would be increased suddenly, both in quantity and intensity, and all the phenomena of "blazing" stars be revealed in the precise order in which we see them. Professor Proctor seems to have had some such idea of space vaguely in his mind when he says, in his "Familiar Essays," "One is invited to believe that the star may have been carried by its proper motions into regions where there is a more uniform distribution of the material whence this orb recruits its fires. It may be that, in the consideration of such causes of variation affecting our sun in long-past ages, a more satisfactory explanation than any yet obtained may be found of the problem geologists found so perplexing,—the former existence of a tropical climate in places within the temperate zone, or even near the arctic regions. Sir John Herschel long since pointed to the variation of the sun as a possible cause of such changes of climate." In confirmation of the view that such changes may be due to the passage of a solar system into or through such a "Gulf Stream" of space, we quote the following from Professor Proctor's "Suns in Flames:" "It is noteworthy that all the stars which have blazed out suddenly, except one, have appeared in a particular region of the heavens,—the zone of the Milky Way (all, too, in one-half of that zone). The single exception is the star in the Northern Crown,

and that star appeared in a region which I have found to be connected with the Milky Way *by a well-marked stream of stars;* not a stream of a few stars scattered here and there, but a stream where thousands of stars are closely aggregated together, though not quite so closely as to form a visible extension of the Milky Way. . . . Now, the Milky Way and the outlying streams of stars connected with it seem to form a region of the stellar universe where fashioning processes are still at work." In just such regions of potential energy should we look for such currents in space, as, on our own earth, the Gulf Stream and the trade-winds, as well as cyclones and other atmospheric movements, find their origin under precisely parallel circumstances, —to wit, the outpour upon and direct precipitation of increased quantities of heat at the tropics or other local centers of such development. The effects of such an increase of quantity and potential in an electrical current are clearly illustrated in the device previously referred to, in which electrolytic decomposition was effected in a pail of water; we find it also in the burning out of the brushes and commutators in dynamo-electric machines and in telegraphic apparatus during thunder-storms and the like. Allowing a solar system a drift through space only equal to that of our own, which has a relatively slow movement, it would traverse such a "Gulf Stream" of space seven hundred thousand miles wide in a single day. But it may not even have passed through; it may merely have grazed the margin of such a current; for the motions of solar systems are not controlled by the same forces as those upon which their electrical energies depend.

Professor Ball, in his chapter on the great heat-wave of 1892, says, "Towards the end of July an extraordinarily high temperature, even for that period of the year, prevailed over a very large part of the North American continent. The so-called heat-wave then seems to have travelled eastward and crossed the Atlantic Ocean; ... a fortnight after the occurrence of unusually great heat in the New World there was a similar experience in the Old World. ... This discussion will at all events enable us to make some reply to the question which has often been asked, as to what was the cause of the great heat-wave. ... It is, however, quite possible that certain changes in progress on the sun may act in a specific manner on our climate. ... It cannot be denied that local, if not general, changes in the sun's temperature must be the accompaniment of the violent disturbances by which our luminary is now and then agitated. It is, indeed, well known that there are occasional outbreaks of solar activity, and that these recur in a periodic manner; it is accordingly not without interest to notice that the present year has been one of the periods of this activity. We are certainly not going so far as to say that any connection has been definitely established between a season of exuberant sun-spots and a season remarkable for excessive warmth; but, as we know that there is a connection between the magnetic condition of the earth and the state of solar activity, it is by no means impossible that climate and sun-spots may also stand in some relationship to each other." These local deviations are doubtless due to planetary positions with reference to the sun, but more general variations must depend upon the constitution of such parts of space as the solar system may occupy; but even then they will be but temporary, since the sun's volume will rapidly expand or contract so as finally to restore the normal emission of solar heat, as will be further explained later on in this work.

There are other causes also, readily conceivable, for such increased electrical action; for instance, in that thickly-peopled region of space, two solar systems adjacent might easily have their exterior planets so related to each other as suddenly, at their points of nearest approach, to cause one or more to direct an abnormally large electrical current into the sun of the adjacent system; this would correspond in electric energy, in fact, to a violent "perturbation" in its orbit by the action of gravity produced by a neighboring planet or system. No reversal of polarity could take place between these planets under these circumstances, any more than between the earth and the moon. In some portions of the Milky Way, doubtless, suns blaze by dozens across the sky at night, and by day as well, to which, in our more solitary skies, we are strangers. Revolving in perfect harmony, perturbations must nevertheless be frequent, and to what limits they may there be confined we shall never know until we realize the extent of these galaxies and the relative contiguity of their solar systems to each other. It is enough to show how such variations may occur; in what particular way they do occur does not affect the question of their origin. Even if such increased energy were to continue by permanently increased planetary action, it is not necessary to suppose that a corresponding permanent increase of light and heat would result on the part of the sun, for its density is such (only one-fourth that of the earth) that, under the tremendous force of its gravity (twenty-seven and one-tenth times that of the earth), its constituents cannot be maintained in solid form, but must be, as before stated, either liquid or gaseous, and perhaps in part both. Now, as it has been computed that the sun, by contraction to its present density, would have evolved its present light and heat for a period of millions of years, it is obvious that any increase in its present volume, without increase of mass, would produce precisely opposite and compensated results, so that the sun could receive from outside sources as much heat as would expand its present volume to that at the initial point of such assumed condensation without increased emission of light and heat. The sun is thus, in effect, a self-compensating machine, and its passage through a region of increased electrical generation would first manifest itself in a vast increase of brilliancy, due to higher incandescence of its hydrogen envelope; this, in turn, would be communicated to the deeper structures of the sun, producing increased volatili-

196   SOURCE AND MODE OF SOLAR ENERGY.

zation and dark absorption bands, and finally to the whole solar mass, expanding its volume in proportion to the heat absorbed. Hence we should see

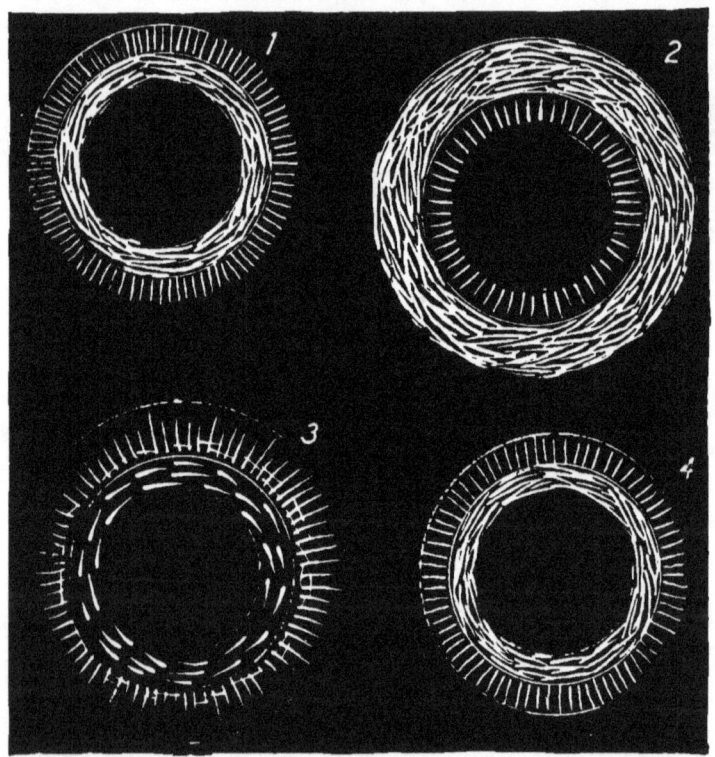

Phenomena of a new or temporary star, a "star in flames."—1, normal state of star, photosphere more highly heated than chromosphere; 2, stage of highest incandescence, chromosphere expanded and more highly heated than photosphere, bright line spectrum; 3, stage of recession, chromosphere diminishing in incandescence, heat acting upon solar core, numerous spots, volatilization of metallic surface, spectrum of dark absorption bands; 4, return to normal state again.

precisely the phenomena that we do see in flaming stars or so-called new stars. We find such compensations all through nature, and it is simply in accordance with her universal laws that they occur. It is a singular circumstance that the catastrophe which is foretold in the biblical record as the termination of all human life on earth, for the present cycle, at least, should be almost literally in accordance with the phenomena characteristic of such an increase of solar energy, and one produced in some such manner. If the temperature of the solar atmosphere were rapidly raised by increased planetary action to a point which would reverse the lines of hydrogen from dark to bright, say to a brightness eight hundred times that of the normal, as in the case of the temporary star cited, though the heat would not, of course, be increased in any such proportion, yet the heavens would be indeed rolled up as a scroll, and all life would be extinguished in a very brief period. But the planets would continue to roll along their orbits, the integrity of the earth's mass would still be intact, and after a few days or weeks the sun would begin to decline in brightness, the volatilized vapors would slowly recede within the solar atmosphere, and the temperature would gradually fall again to its normal, leaving, however, a lifeless world to roll on its way henceforth, but as bright and cheerful in all its possibilities, when the former conditions had gradually become restored, as before. Perhaps some distant astronomer in the neighborhood of Sirius—if we shall have travelled so far away by that time—might send a note to the morning papers to announce that the temporary star near Alpha Centauri had again receded to the tenth magnitude. In due time—perhaps a thousand years—all would be ready for a new development of life, and the cycle would continue as before. Perchance, too, in some deep abyss, or buried far beneath the surface, some germs of life might still continue to exist; and from these, like the seeds resurrected from buried mummies, a new life might again begin, guided along once more through vast ages in a progressive ascent from development to development until, in some new and strange forms, the higher types of life might again appear. To these there would indeed be revealed a new heaven and a new earth. Who knows how many such cycles of life may have come and gone on earth, in which, like the dwellers of Jerusalem, new peoples have built new cities, one above another, upon the unknown graves of the past? In the words of Tennyson,—

"A wondrous eft was of old the Lord and Master of earth,
For him did his high sun flame, and his river billowing ran,
And he felt himself in his force to be Nature's crowning race.
As nine months go to the shaping an infant ripe for his birth,
So many a million of ages have gone to the making man:
He now is first, but is he the last?"

Whatever the coming, the progress, or the going of life on earth, the course of our solar system will go on the same, the processes of creation unchanged and her mechanism unimpaired. It is obvious that no such conditions could prevail in the return to unorganizable chaos which must be the consequence of any possible planetary collisions in space. No conceivable process of creation could return a system disrupted into meteorites to an operative solar system again. Even the nebular hypothesis contemplates nothing of that sort as, by the wildest conjecture, ever possible. But with us the danger is far distant. Professor Proctor says, in his article "Suns in Flames," "As Sir William Herschel long since pointed out, we can recognize in various parts of the heavens various stages of development, and chief among the regions where as yet nature's work seems incomplete is the Galactic zone,—especially that half of it where the Milky Way consists of irregular streams and clouds of stellar light. As there is no reason for believing that our sun belongs to this part of the galaxy, but, on the contrary, good ground for considering that he belongs to the class of insulated stars, few of which have shown signs of irregular variation, while none have ever blazed suddenly out with many hundred times their former lustre, we may fairly infer a very high degree of probability in favor of the belief that, for many ages still to come, the sun will continue steadily to discharge his duties as fire, light, and life of the solar system." The passage of our system through gradually changing regions of space, as contrasted with streams or vortices, could not affect our sun's light even temporarily, as the contraction and expansion of its volume would fully compensate for any such gradual or partial variation, and, by position, he is far from likely to pass into any of those whirlpools or torrents of space which seem to mark at irregular intervals the region of the irregularly variable stars.

Allied in appearance to such stars which suddenly flame out in space, but totally different in reality, are comets. These strangers to our own system have excited the wonder and astonishment of mankind from the earliest ages. They seem to defy all rules and all explanation; but, when properly examined, they will fall inevitably into the general scheme of the source and mode of solar energy which we have endeavored to present. These bodies enter our solar system from without. Appleton's Cyclopædia says, "Schiaparelli, to whom the discovery is in part due, considers the meteors to be dispersed portions of the comet's original substance,—that is, of the substance with which *the comet entered the solar domain.*" Professor Proctor, "Meteoric Astronomy," says, "A word or two may be permitted on the question of the condition of *comets freshly arriving on the scene of the solar system.* It is assumed sometimes that the train of meteors already exists when the comet *first comes within the solar domain.*" In the "Romance of Astronomy" (R. Kalley Miller, M.A.) it is said, "In a sort of debatable territory between our own solar system and the infinite stellar universe around we come upon these erratic and anomalous bodies —the comets; some of which have accidentally become permanent attendants upon our sun; others have only paid it a single casual visit in the course of their wanderings through space, and are not likely again to come within the range of its attracting influence; while countless millions are doubtless scattered throughout the realms of the infinite, whose existence will never be revealed to human ken at all." Professor Helmholtz, in fact (see addendum to his lecture on the origin of the planetary system), advanced the idea in a speculative way, that our terrestrial life might have had its origin in one of these meteoric bodies by the "transmission of organisms through space." In Professor Proctor's article on comets ("Mysteries of Time and Space") he says, "The paths followed by comets show no resemblance either to the planetary orbits or to each other. Here we see a comet travelling in a path of moderate extent and not very eccentric; then another which rushes from a distance of two or three thousand millions of miles, approaches the sun with ever-increasing velocity until nearer to him than parts of his own corona (as seen in eclipses), sweeps around him with inconceivable rapidity, and makes off again to where the aphelion of its orbit lies far out in space beyond the most distant known planet,—Neptune. Some comets travel in a direct, some in a retrograde path; a few near the plane of the earth's orbit, many in planes showing every variety of inclination. Some comets regularly return after intervals of a few years; some after hundreds of years; others are only seen once or twice, and then unaccountably vanish; and not a few show by the paths they follow that they have come from interstellar space to pay our system but a single visit, passing out again to traverse we know not what other systems or regions. . . . When we have said that these objects obey the law of gravity, we have mentioned the only circumstance—as it would appear—in which they conform to the relations observed in terrestrial and planetary arrangements. And even this law—the widest yet revealed to man—they seem to obey half unwillingly. We see the head of a comet tracing out systematically enough its proper orbit, while the comet's tail is all unruly and disobedient. . . . The fact, then, is demonstrated that two of the meteor streams encountered by the earth are so far associated with two comets as to travel on the same orbits. We may not unsafely infer that all the meteor systems are in like manner associated with other comets. Nor is it very rash to assume that all comets are in like manner associated with meteor systems."

Concerning the influence of gravitation of the planets, the same author says (" Meteoric Astronomy"), "Now, the circumstances under which a comet approaching the sun on a parabolic or hyperbolic orbit can be thus affected must be regarded as exceptional. The planet's influence must, in the first place, be very energetically exercised; in other words, the arriving comet must pass very close to the planet, for under any other circumstances the sun's influence so enormously outvies the planet's that the figure of the cometic orbit would be very little affected. Moreover, the planet's attraction must produce an important balance of retardation. The planet will inevitably accelerate the comet up to a certain point, and afterwards will retard it; the latter influence must greatly exceed the former. To show how greatly the comet must be retarded, it is only necessary to mention that the actual velocity of the November meteors when they cross the orbit of Uranus is less than one-third of the velocity with which Uranus himself travels, but their velocity at the same distance from the sun, when they were approaching him from some distant stellar domain, exceeded the velocity of Uranus in his orbit in the proportion of about seven to five. . . . It follows, not merely as a probable inference, but, I think, as a demonstrated conclusion, that if the November meteors came originally into our system as a comet travelling sunward from infinity, then either that comet was very compact or else Uranus captured only a small portion of the comet, the remaining portions moving thenceforth on orbits wholly different from the path of the November meteors. . . . No other planet than Uranus can have brought about the subjection of this comet to solar rule." In his article on comets he says, "It may be well here to consider a case in which some active force (other than gravity) exerted by the sun seems to have brought the destruction of a comet, or at least to have broken up the comet into unrecognizable fragments." He refers to Biela's comet, with an orbital period of six and two-thirds years, and a path which was found to approach very near to the path of the earth. In 1832 the comet crossed the earth's track several weeks before the arrival of the earth at the same point without appreciable interference. On its second return, in 1845–46, it was found to be divided into two comets travelling side by side; in

1852 they reappeared, still divided, and gradually diverging from each other. Since then they have never reappeared, though diligently sought for at every period. Professor Proctor adds, "It has been seen again, though not as a comet; nay, the occasion on which it was seen in the way referred to was predicted, and the prediction fulfilled, even in details. For a full account of its reappearance—as a meteor stream—I refer the reader to my essay on Biela's comet in 'Familiar Science Studies.'"

In Miller's "Romance of Astronomy" we read, "Encke's comet, which possesses the smallest orbit of any connected with our system, is sensibly drawing nearer and nearer to the sun at every revolution." In Professor Proctor's "Cometic Mysteries," the author says, "We hear it stated that the nucleus of a comet is made up of meteoric stones (Professor P. G. Tait says—for unknown reasons—that they resemble 'paving stones or even bricks') as confidently as though the earth had at some time passed through the nucleus of a comet, and some of our streets were now paved with stones which had fallen to the earth on such an occasion. As a matter of fact, all that has yet been proved is that meteoric bodies follow in the track (which is very different from the tail) of some known comets, and that probably all comets are followed by trains of meteors. These may have come out of the head or nucleus in some way as yet unexplained; but it is by no means certain that they have done so, and it is by many astronomers regarded as more than doubtful. The most important point to be noticed in the behavior of large comets as they approach the sun is, that usually the side of the coma which lies towards the sun is the scene of intense disturbance. Streams of luminous matter seem to rise continually towards the sun, attaining a certain distance from the head, when, assuming a cloud-like appearance, they seem to form an envelope around the nucleus. This envelope gradually increases its distance from the sun, growing fainter and larger, while within it the process is repeated and a new envelope is formed. This, in turn, ascends from the nucleus, expanding as it does so, while within it a new envelope is formed. Meanwhile the first one formed has grown fainter, perhaps has disappeared. But sometimes the process goes on so rapidly (a day or two sufficing for the formation of a complete new envelope) that several envelopes will be seen at the same time,—the outermost faintest, the innermost most irregular in shape and most varied in brightness, while the envelope or envelopes between are the best developed and most regular. The matter raised up in these envelopes seems to have undergone a certain change of character, causing it no longer to obey the sun's attractive influence, but to experience a strong repulsive action from him, whereby it is apparently swept away with great rapidity to form the tail. 'It flows past the nucleus,' says Dr. Huggins, 'on all sides, still ever expanding and shooting backward until a tail is formed in the direction opposite to the sun. This tail is usually curved, though sometimes rays or extra tails sensibly straight are also seen.'"

In "The Sun as a Perpetual Machine," Professor Proctor says, "Take, again, the phenomena of comets, which still remain among the greatest of nature's mysteries. We have reason to believe . . . that the nucleus of a comet consists of an aggregation of stones similar to meteorites. Adopting this view, and assuming that these stones have absorbed somewhere gases to the amount of six times their volume (taken at atmospheric pressure), we may ask, What will be the effect of such a mass of stones advancing towards the sun at a velocity reaching in perihelion the prodigious rate of three hundred and sixty-six miles per second (as observed in the comet of 1843), being twenty-three times our orbital rate of motion?" Professor Ball says, "One of the most important results of the great shower of 1866 was the demonstration that the swarm of little bodies to which that shower owed its origin was connected with a comet. The swarm was found, in fact, to follow the exact track which the comet pursued around the sun. . . . Of this connection between the cometary orbits and revolving swarms of meteors many other instances could be cited. I may refer to the remarkable lists published by the British Association, in which, beside the name of the comet or the designation which astronomers had affixed to it, the meteoric swarm with which the comet is associated is also given. . . . On these grounds it appears to be perfectly certain that the origin of the shooting stars which appear in swarms cannot be disassociated from the origin of the comets by which those swarms are accompanied." The author makes a distinction between such ordinary shooting stars and meteorites, and attributes the appearance of the latter on earth to masses thrown forth from some volcano *somewhere*, but this has nothing to do with the special phenomena to be interpreted. It may be said, however, that the presence of olefiant gas as one of the occluded gases in a meteorite (four and fifty-five-hundredths per cent., as stated by Professor Proctor, in his article "The Sun as a Perpetual Machine"), and the remarkable fact, stated in the article "Spectrum Analysis" in Appleton's Cyclopædia, that, in Winnecke's comet of 1868, "the bands agree in position with those obtained as the spectrum of carbon, by passing the electric spark through olefiant gas," would lead one to consider a cometic origin, for this particular meteorite at least, to be highly probable. Professor Ball further says, "There have been several instances in which a comet has approached so close to a planet that the attraction between the two bodies must have had significant influence on the planet, if the cometary mass had been at all comparable with that of the more robust body. The most celebrated instance is presented in the case of Lexell's comet, which happened to cross the track of Jupiter. The effect upon this body was so overwhelming that it was wrenched from its original path and started afresh along a wholly different track." The same writer, speaking of the tails of comets, says, "I have no intention to discuss here the vexed question of the tails of comets. I do not now inquire whether the repulsion by which the tail is produced be due to the intense radiation from the sun, or to electricity, or to some other agent. It is sufficient for our present purpose to note that, even if the tails of comets do gravitate towards the sun, the attraction is obscured by a more powerful repulsive force. . . . Nor do the directions in which the comets move exhibit any conformity; some move round the sun in one direction, some move in the opposite direction. Even the planes which contain the orbits of the comets are totally different from each other. Instead of being inclined at only a very few degrees to their mean position, the planes of the comets hardly follow any common law; they are inclined at all sorts of directions. In no respect do the comets obey those principles which are necessary to prevent constitutional disorder in the planetary system. . . . Now, all we have hitherto seen with regard to comets tends to show that the masses of comets are extremely small. Attempts have been made to measure them, but have always failed, because the scales in which we have attempted to weigh them have been too coarse to weigh anything of the almost spiritual texture of a comet. It is unnecessary to go as far as some have done, and to say that the weight of a large comet may be only a few pounds or a few ounces. It might be more reasonable to suppose that the weight of a large comet was thousands of tons, though even thousands of tons would be far too small a weight to admit of being measured by the very coarse balance which is at our disposal." In the chapter "Visitors from the Sky," the same author says, "As such a comet in its progress across the heavens passes between us and the stars, those stars are often seen twinkling brilliantly right through the many thousand miles of cometary matter which their rays have to traverse. The lightest haze in our atmosphere would suffice to extinguish the faint gleam of these small stars; indeed, a few feet of mist would have more power of obstructing the stellar light than cometary material scores of thousands of miles thick. It is true that the central portions of many of these comets often exhibit much greater density than is found in the exterior regions; still, in the great majority of such objects there is no opacity, even in the densest part, sufficient to put out a star. In the case of the more splendid bodies of this description, it may be supposed that the matter is somewhat more densely aggregated as well as more voluminous; still, however, it will be remembered that the great comet of 1858 passed over Arcturus, and that the star was seen shining brilliantly, notwithstanding the interposition of a cometary curtain millions of miles in thickness. So far as I know, no case is known in which the nucleus of a really bright and great comet has been witnessed in the act of passage over a considerable star. It would indeed be extremely interesting to ascertain whether in such case the star experienced any considerable diminution in its lustre."

CHAPTER VIII.

THE PHENOMENA OF COMETS.

From the extracts thus cited we may form a fairly clear idea of the phenomena which comets present, and these facts represent about all that we know of these mysterious objects. They approach the sun in a nearly radial direction, thus cutting the planetary orbits transversely. They approach the sun from all directions and at all angles, without reference to the common plane in which all the planetary orbits lie. They have no rotation on their own axes, as the planets have, but, like an aggregated mass of meteorites or cosmical dust, rush inward from the exterior realms of space, so that their course is diametrically opposite that of the planets and the other cosmical bodies which constitute our solar system. Such a body as a comet, in fact, would present in its approach to our solar system very much the phenomena of an approaching exterior sun, corresponding far more closely in appearance and behavior to our own sun than to any of the planets. Such a body could not generate positive electricity, as the planets do, but, on the contrary, must have an electrosphere of negative, or at least neutral, polarity. On its approach to our planetary system the batteries of all the planets would be at once turned upon the intruder, and it would be rapidly thrown into the same state of active electrical polarity as the sun. The aqueous vapor condensed around its nucleus by gravity in its approach through space, or buried among the meteoric particles constituting the

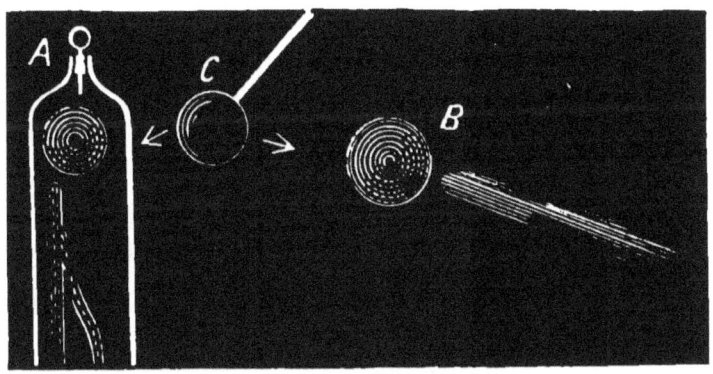

Repulsion of glow in partial vacuum compared with phenomena of sun and comet.—*C*, charged electrical conductor; *A*, electrical discharge in partial vacuum, repelled by like electricity of *C*; *B*, Henry's comet, *C* representing the sun.

comet, would be necessarily decomposed into its constituent gases, just as in the case of the sun, by the positive electrical currents from the planetary electrospheres, and the disassociated hydrogen would form the negative electrosphere of the comet, glowing with its own luminosity, by gaseous incandescence. We should then observe, during its continued approach to the sun, phenomena similar to those which we might expect to manifest themselves during the approach of a minute solar body towards the sun, characterized by a rapid increase of velocity, due to attraction of gravity, and tremendous mutual repulsion between the solar and cometic electrospheres. We should see the luminous hydrogen and associated gases boiling upward, and thence drawn forward from the nucleus by the combined gravity of the sun's mass, that of the planetary masses, and the opposite polarity of the planetary electrospheres, while they would be, at the same time, repelled backward by the enormous repulsive force of the negative electrosphere of the sun. As a result, we should find these gases in a state of ebullition, forced forward under great excitement and disturbance, boiling, eddying about, driven to and fro in all directions until the sun's repulsive force had overcome the different attractions, when these luminous clouds or envelopes would be swept swiftly off to the rear, as by a powerful current of wind, around the margins of the nucleus, and they would be seen to stream backward from the sun as an elongated envelope or tail. New volumes of gas would pour to the front, attracted from deeper depths, and these, on reaching the cometary electrosphere, would be again repelled by the solar activity and driven to the rear, while the gases thus driven backward, themselves similarly electrified, would mutually repel each other as they streamed backward around the margins of the nucleus.

Let us now see what these gases are: if they are such as appear in the sun's electrosphere, we will know that such must be their action; if, on the contrary, they are such as appear in planetary electrospheres, we will find any such attempted explanation to be a failure. Quoting largely from Dr. Huggins, Professor Proctor, in his "Cometic

Mysteries," says, "The spectrum of the brightest comet of that year was partly continuous, and on this continuous spectrum many of the well-known Fraunhofer lines could be traced. This made it certain that part of the comet's light was reflected sunlight, though Dr. Huggins considers also that a part of the continuous spectrum of every comet is due to inherent light. On this point some doubt may be permitted. It is one thing for special bands to show themselves, for some substances may become self-luminous under special conditions at very moderate temperatures; it is quite another thing that the solid parts of a comet's substance should become incandescent. I venture to express my opinion that this can scarcely happen, except in the case of comets which approach very near to the sun. Besides the continuous spectrum with dark lines, the *photograph showed also a spectrum of bright lines.* 'These lines,' says Dr. Huggins, 'possessed extreme interest, for there was certainly contained within this hieroglyphic writing some new information. A discussion of the position of these new lines showed them to be undoubtedly the same lines which appear in certain compounds of carbon. Not long before Professors Liveing and Dewar had found from their laboratory experiments that these lines are only present when nitrogen is also present, and that they indicate a nitrogen compound of carbon,—namely, cyanogen. *Two other bright groups were also seen in the photograph, confirming the presence of hydrogen,*—carbon and nitrogen.' It is worthy of notice that only a few days later Dr. H. Draper succeeded in obtaining a photograph of the same comet's spectrum. It appeared to him to confirm Dr. Huggins's statements, except only that the dark Fraunhofer lines were not visible, the photograph having probably been taken under less favorable conditions. . . . But the latest comet has brought with it fresh news. Its spectrum is not like that given by the comets we are considering. The bright lines of sodium are seen in it, and also other bright lines and groups of lines which have not yet been shown to be identical with any belonging to the hydrocarbon groups, but probably are so. . . . The cyanogen groups are not seen. . . . But it is manifest that *this comet underwent important changes.* . . . In April was found simply a faint continuous spectrum; in May the three bands associated with carbon were present, though faint, while there was no trace whatever of the sodium band. On the contrary, in June the nucleus of the comet gave a very strong and extended continuous spectrum with an excessively strong bright line in the orange-yellow identical with the well-known double sodium line of the solar spectrum. On this . . . it is necessary to conclude that during the last fortnight of May the spectrum of Wells's comet had changed in a manner of which the history of science furnishes no precedent."

It should be observed that the elements carbon and hydrogen closely resemble each other, not only in their multifarious chemical affinities and reactions, but in their electric polarities, and the hydrocarbon compounds, like their constituents, carbon and hydrogen, are electrically similar to each other, an example of this similarity of the elements being found in the identical action of the carbon arc and hydrogen envelope in the heating and lighting experiments with electrical currents hereinbefore described.

We have already seen that carbon follows quite a different law from the other concrete elements, in the fact that its electrical resistance diminishes as the temperature rises; it also differs widely from the other solid elements in its *atomic heat*, which has a value much less than one-half the mean constant, which is 6.4. Of this matter of specific heat, Professor Fownes, in his work on chemistry (Bridges' edition), says, " Dulong and Petit observed in the course of their investigation a most remarkable circumstance. If the specific heats of bodies be computed upon equal weights, numbers are obtained all different and exhibiting no simple relations among themselves; but if, instead of equal weights, quantities be taken in the proportion of the atomic weights, an almost perfect coincidence in the numbers will be observed, showing that some exceedingly intimate connection must exist between the relations of bodies to heat and their chemical nature; and when the circumstance is taken into view that relations of even a still closer kind link together chemical and electrical phenomena, it is not too much to expect that ere long some law may be discovered far more general than any with which we are yet acquainted . . . .

Nevertheless, this law must not be understood as perfectly general, for there are three elements—namely, carbon, boron, and silicon" [these form a single group of elements in chemical classification]—"which exhibit decided exceptions to it."

Organic chemistry is substantially based upon the almost infinitely interchanging relations among carbon-hydrogen radicals, supplemented by a few other elements. According to Professor Fownes, "Organic chemistry is in fact the chemistry of carbon compounds." The position of carbon among the elements is something like that of camphor among the oils, the latter being a volatile oil, but concrete in form. With a concrete element having the peculiar character of carbon we can well understand its universal chemical and electrical relationship with gaseous hydrogen in the grandest operations of nature.

Cyanogen is an electrically similar compound of carbon with the addition of nitrogen. Of these elements it will be seen that nitrogen and hydrogen are found to exist also in the gaseous nebulæ, and *with the probable addition there of oxygen;* but in comets the quota of active oxygen must be sought for in the correlated planetary, and not in the cometic, atmospheres, as is the case with the sun. Of the presence of the vapor of carbon in comets Professor Ball says, "This is a very singular fact, when it is remembered that carbon is one of the substances essentially associated with life in the forms in which we know it." Professor Huggins says, "Since that time the light from some twenty comets has been examined by different observers. The general close agreement in all cases, notwithstanding some small divergencies, of the bright bands in the cometary light with those seen in the spectrum of hydrocarbons justifies us fully in ascribing the original light of these comets to matter which contains carbon *in combination with hydrogen*."

We may learn something further of the constitution of comets, perhaps, by considering the chemical reactions which their spectra seem to indicate. The following extract is from a recent article on the manufacture of illuminating gas: " Ammonia contains 82.35 parts of nitrogen and 17.65 of hydrogen. It is not produced by a direct combination, for nitrogen can be caught and wedded only by a hot and skilful wooing. In the gas retort, at a temperature of 2200 degrees and in the presence of lime, *soda*, or potash, it will combine with carbon and form cyanogen, and then further combine with the alkali to form a cyanide. There is steam in the retort, and, as nearly as the gas chemists can make out, the nitrogen promptly divorces itself, gives up the carbon to the oxygen of the steam, and, taking the hydrogen to itself, becomes, for the time at least, a fixed, if volatile, substance, but ever ready to enter into new alliances." It will be remembered that in the comets examined by Professors Huggins and Draper the spectroscope revealed both cyanogen and the double line of sodium. The function of the sodium is readily understood, as by its presence it enables the nitrogen in the cometic atmosphere to combine with a part of the carbon of the gaseous hydrocarbons which constitute this atmosphere, and thus produce the cyanogen. But to effect this combination requires in the retort a temperature of 2200 degrees. If the combining temperature around the nucleus of a comet is the same, it will show that the temperature of this comet's nucleus must be very high, and, while many times less than that of the sun's photosphere, it still clearly illustrates the powerful character of the impact of the planetary electrical currents upon the comet, and its tremendous repulsion by the similarly electrified solar electrosphere. The second one of the above reactions, that from cyanogen to ammonia, is due to the steam or aqueous vapor in the retort. But in the case of the comet all the aqueous vapor and its constituent oxygen have disappeared by electrolytic decomposition long before the combining temperature of cyanogen has been reached; so that the sodium, the hydrocarbons, and the cyanogen alone appear, and the oxygen compounds are missing. But on the reversal of polarity of this comet by contact with a planetary electrosphere, should such ever occur, and its consequent assumption of positive electricity, the oxygen would again appear, and, if the temperature had not yet receded below that of the reaction which produces ammoniacal vapors, we might expect, should a fragment of this comet enter our atmosphere as a meteorite, to find ammonia as well as sodium as a constituent thereof; otherwise the ammonia would be replaced by carbonic oxide and carbonic acid, by the action of oxygen upon the hydrocarbons, and water by the action of oxygen upon the hydrogen of the same, at much lower temperatures than would suffice for the generation of ammonia. The cyanogen would then perhaps remain as cyanide of sodium, unless decomposed by contact with the meteoric metallic iron at a high temperature, as occurs in the operation known in the arts as "case-hardening." The presence of microscopic diamonds in meteorites may be accounted for by a somewhat similar reducing reaction under heat and the active force of the planetary and solar voltaic arc.

In the popular view comets are always associated with tails, but, in fact, comets without tails are far more numerous than those to which these appendages pertain; the tails, when such exist, are the direct result of the repulsive energy of the solar electrosphere, and are only manifested when their proximity to the sun has aroused sufficient activity to swiftly sweep backward from the sun with inconceivable velocity the gaseous matter concentrated in and around the nucleus. As these tails owe their formation to the sun's repulsive energy, they must always extend radially outward from the sun, and by the self-repulsive energy of the diverse constituents of the tails themselves these will be broken occasionally into two, four, or six lateral strands, and (possibly by the attraction of the different planetary electrospheres) curvatures may be apparent along the sweep of the comets' tails corresponding, in effect, with perturbations produced by gravity in the orbit of the nucleus. Of these various phenomena, Professor Proctor, in his article on comets, says, "A very large number of comets have no visible tails. When first seen in the telescope a comet usually presents a small, round disk of hazy light, somewhat brighter near the center. As the comet approaches the sun the disk lengthens, and, if the comet is to be a tailed one, traces begin to be observed of a streakiness in the comet's light. Gradually a tail is formed, which is turned always from the sun. The tail grows brighter and larger, and the head becomes developed into a coma surrounding a distinctly marked nucleus. Presently the comet is lost to view through its near approach to the sun; but after a while it is again seen, sometimes wonderfully changed in aspect through the effects of solar heat. Some comets are brighter and more striking after passing their point of nearest approach to the sun than before; others are quite shorn of their splendor when they reappear." This change of aspect is not due to solar heat, but to the energetic repulsion of the solar electrosphere. The force of gravity irresistibly impels the comet forward to the sun's electrical vortex, and the change of aspect is due to the repulsion of its entire stock of free gaseous matter into space in case its supply is small, or to its increased development and pouring forth in case the supply is large. It is like the volatilization by a heated atmosphere of ammoniacal gas, for instance, absorbed in water. The ebullition is vastly increased by the heat, but if the entire stock of ammonia has been driven off in its passage through the heated medium, it will emerge with the residual water quiescent; otherwise, in a state of increased agitation.

The same author, in " Cometic Mysteries," says, " Repulsion of the cometary matter could only take place if this matter, after it has been driven off from the nucleus, and the sun *have both high electric potentials of the same kind.*" His further guess, however, that it is analogous to the aurora, is wide of the mark; it is due, in fact, to the mutual repulsion of their similar negative electrospheres, the cometic electrosphere, however, being so much smaller than that of the sun that the latter shows no appreciable disturbance, as is the case, under similar circumstances, with the electrospheres of the earth and moon. In the article last quoted it is said, " There is a dark space immediately behind the nucleus,—that is, where the nucleus, if solid, would throw its shadow if there were matter to receive the light all round so that the shadow could be seen." This presents, it is stated, a great difficulty. The author, by a happy guess,—almost an inspiration, in fact, of which this splendid writer and observer was so full,—suggests in a foot-note a possible explanation, which, while not in itself correct, suggests an analogous process very like what we actually see. " If the particles forming the envelopes are minute flat bodies, and if anything in the circumstances under which these particles are driven off into the tail causes them to' always so arrange themselves that the planes in which they severally lie pass through the axis of the tail (which, if the tail is an electrical phenomenon, might very well happen), then we should find the region behind the nucleus very dark or almost black, for the particles in the direction of the line of sight there would be turned edgewise towards us, whereas those on either side or in the prolongation of the envelopes would turn their faces towards the observer." As a matter of fact, the envelope streaming backward from the nucleus forms a hollow tube, the opposite sides of which exhibit the same mutual repulsion as both exhibit towards the sun; hence the phenomenon would be similar to that exhibited by blowing into a closed bag of porous material covered with wisps of cotton, for example, and the gases, in addition to their rush backward from the sun, would also exhibit a radial rush outward from the longitudinal axis of the tail. This is what we actually observe, and sufficiently accounts for the phenomenon, be it altogether or only partially real, and not merely, as that author thinks it may be, apparent. It is said, in the same article, that "Bredichen has shown that where there are three tails to a comet their forms correspond with the theory that the envelopes raised from the head are principally formed of hydrogen, carbon, and iron; but this . . . seems open at present to considerable doubt." At all events, these separate tails are self-repulsive, or they would be merged into each other by the sun's repulsive energy; in fact, they occupy the resultant of the direction produced by the line of the sun's repulsion and those of their own mutually repellent force,—that is to say, radial or divergent.

It must not be supposed that these tails are of insignificant proportions. "When we see the tail of a comet occupying a volume thousands of times greater than that of the sun itself, the question naturally suggests itself, 'How does it happen that so vast a body can sweep through the solar system without deranging the motion of every planet?' Conceding even an extreme tenuity to the substance composing so vast a volume, one would still expect its mass to be tremendous. For instance, if we supposed the whole mass of the tail of the comet of 1843 to consist of hydrogen gas (the lightest substance known to us), yet even then the mass of the tail would have largely exceeded that of the sun. Every planet would have been dragged from its orbit by so vast a mass passing so near. We know, on the contrary, that no such effects were produced. The length of our year did not change by a single second. . . . Thus we are forced to admit that the actual substance of the comet was inconceivably rare. . . . From what we have already seen, it will be manifest that the formation of comets' tails is a process of a very marvellous nature, apparently involving forces other than those with which we are acquainted. The tail, ninety million miles in length, which was seen stretching from the head of Newton's comet nearly along the path which the retreating comet had to traverse, must, it would seem, have been formed by some force far more active than the force of gravity. The distance traversed by the comet in the last *four weeks* of its approach to the sun under gravity was no greater than that over which the matter of the tail, seen after the comet had circled around the sun, *had been carried in a few hours.* Yet we have no other evidence of any repulsive force at all being exerted by the sun,—at least no evidence which can be regarded as demonstrative, —and still less have we any evidence of a repulsive force exceeding in energy the sun's attracting power." (Proctor.)

CHAPTER IX.

INTERPRETATION OF COMETIC PHENOMENA.

Now, curiously enough, we have in constant use in our laboratories a little instrument called the electroscope, in which we have manifested very clearly a repulsive force exceeding in energy the earth's attracting power, and very greatly exceed-

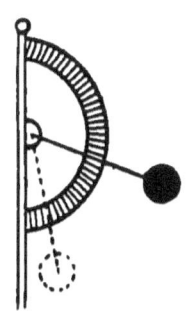

Electroscope, showing repulsion of pithball from charged conductor.

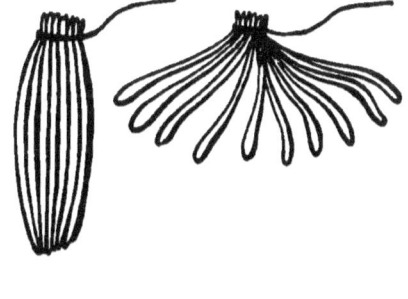

Bundle of straws unelectrified, and afterwards suddenly forced asunder by electricity.

ing it. It is described in "Electricity in the Service of Man" as follows: "If we rub a large glass rod with a silk pad, we observe that it will attract light bodies, then, after contact, repel them. During the process we may notice a peculiar noise, and if the experiment be carried out in the dark we may further notice sparks passing between the rod and the rubber, and also that the rod becomes luminous. If we suspend a pith-ball by means of a silk thread, on bringing the rubbed rod near the pith-ball it will move towards the rod, touch it, and then be repelled. If the glass rod be again brought near the pith-ball, it will move away from the glass rod, and continue to be repelled until it has been touched by some other body. . . . In order to ascertain whether electricity is communicated by electrified bodies to non-electrified bodies when brought into contact, let us suspend two pith-balls from the same point of support by threads of uniform silk, and touch the pith-balls with the rubbed glass rod. The balls fly from the rod and also from one another. On bringing near them a third pith-ball or any other light body, we find that, though they repel one another, they are attracted by the light body, showing that they have become electrified by contact with the rubbed glass rod. From this we conclude that an unelectrified body may be electrified by contact with an electrified body, and also that there is repulsion after contact. There is *mutual repulsion between two electrified bodies*, but there is attraction between a single electrified body and one that is unelectrified." The mutual repulsion of these pith-balls is the exact measure of the strength of electrification. Hung side by side to the knob of a prime conductor of an electrical machine, the mutual repulsion of the similar electrospheres of these pith-balls drives them apart against the earth's gravity and holds them extended, if the electrical tension be sufficient, to their widest limit of divergence. It is, in effect, precisely similar to

INTERPRETATION OF COMETIC PHENOMENA. 227

the action of the solar and cometic electrospheres (see illustration in a previous chapter, page 124), each being similarly electrified and communicating with the other across a space which, as before stated, is freely traversable by electric currents without appreciable resistance. That such electrospheres are flaming with heat does not interfere with such self-repellent action; in fact, it intensifies it. In Professor Tyndall's " Lessons in Electricity" we read, "*Flames* and glowing embers act like points; they also *rapidly discharge electricity.* The electricity escaping from a point or flame renders the air self-repulsive. The consequence is that when the hand is placed over a point mounted on the prime conductor of a machine in good action a cold blast is distinctly felt. . . . Wilson moved bodies by its action, Faraday caused it to depress the surface of a liquid, Hamilton employed the reaction of the electric wind to make pointed wires rotate. The 'wind' was also found to promote evaporation."

Let us now apply these principles to the tails of comets. If we conceive the sun and comet to be analogous to our pith-balls, one enormously larger than the other, however, and hung by vaporous conducting cords from the combined generating planetary electrospheres, both sun and cometic nucleus surrounded each by a vaporous envelope, and suspended so that they will hang from parallel cords, say a dozen million miles apart, and with no currents of electricity as yet in operation, we will find that the sun and comet will be simply attracted towards each other by the force of gravity, so that their suspending cords will converge. If the planetary electrical machines now commence their rotations, and currents of electricity begin to pass in quantity and intensity like those which pass between the earth and the sun, both the solar and cometic pith-balls will become similarly electrified, and their gaseous atmospheres, instead of drawing towards each other, will become luminous and self-repulsive. The amosphere which surrounds the cometic pith-ball, by reason of its great tenuity, will be driven backward with extreme velocity, while the solar pith-ball electrosphere will be so little affected that its repulsion will be imperceptible. All the gaseous matter, however, of the smaller pith-ball will be forced off in a direction opposite that of the larger one, and this repulsive energy will even carry the pith-balls apart, causing the suspending cords to widely diverge from each other, while the force of gravity of the earth tends to bring them nearer together. If the gravity of the larger pith-ball, however, was equal, relatively, to that of the sun, the result would be that the solid pith-balls would be mutually attracted by gravitation and only the electrified atmospheres, would be mutually repelled. This experiment would present phenomena similar to those we are now considering. (See illustration, page 211.)

In describing Newton's comet, with a tail ninety million miles long projected backward both from the sun and the comet, when it disappeared in the light of the sun, and exhibiting a similar tail, also ninety million miles long, when, less than four days afterwards, it reappeared from behind the sun, but with the tail now directed forward from the comet, but in both cases extended radially outward from the sun, it is obvious that this whole tail must have made a sweeping change of direction of nearly one hundred and eighty degrees upon the nucleus as its center. Professor Proctor says, "As Sir John Herschel remarks, we cannot look on the tail of a comet as something whirled round like a stick as the comet circles around its perihelion sweep. The tail with which the comet reappeared must have been an entirely new formation." It is true that a comet's tail cannot be conceived of as being whirled round like a stick, but we can very readily conceive of it as something like a flame composed of incandescent gases, and it may very easily be *blown* round a stick; and this is precisely what must happen in the case of a comet. Construct, for experiment, a little apparatus consisting of a blow-pipe adapted to deliver a current of air between two horizontal metal disks, say an eighth of an inch apart, one perforated at the center to admit the nozzle of the blow-pipe. By directing a constant current of air through the latter, it will be deflected so as to blow radially outward in all directions and in the same plane. Now take a stick with a flame on the end of it, or a lighted candle, and with it approach this center of repellent energy in the plane of the space between the disks and along an ellipse representing the orbit of a comet. As the flame approaches the improvised solar center it will be driven backward from the wick of the candle almost along the line of its approach, and as it

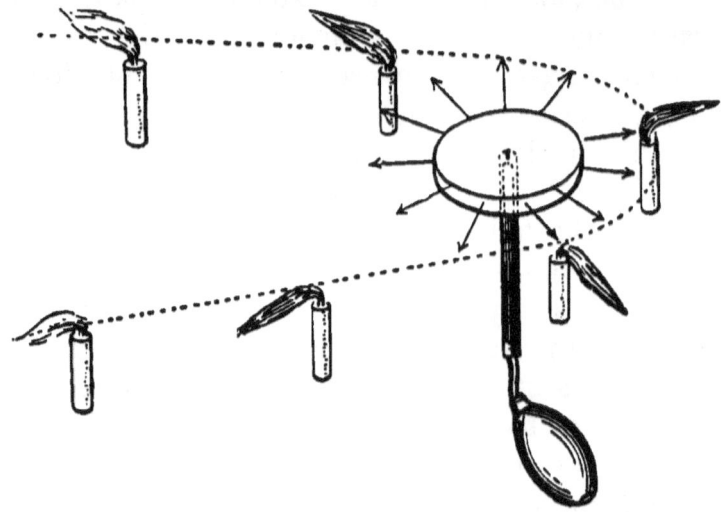

Mechanical device illustrating repulsion by the solar electrosphere of a comet's tail.

passes around the center it will be constantly blown outward in a radial direction until, when it recedes after perihelion, the flame will be seen pointed almost directly ahead. At all times the direction of the flame will lie along the radial lines prolonged outward from the center through the wick of the candle, and it will not be a new flame generated at every change of its direction, but the same flame constantly forced outward by the repulsive force of the central atmosphere in this case or the solar electrosphere in the case of the sun. This experiment is an accurate and conclusive exhibit of the phenomena of solar repulsion in its action upon the tail of a comet. It is analogous in principle to the repulsion of the pith-balls and the electric wind and (in application) to the phenomena presented by comets in their movements to, around, and from the sun. This repulsion is not operative in effect against the wick of the candle,—that is to say, it is not the repulsion of the nucleus which determines the direction of the tail, but the repulsion by direct outblow of the sun, so to speak, upon the incandescent gases of the tail itself. This fact clearly demonstrates that the repulsion of like electrospheres is the cause of the phenomenon, and, when once understood, the process is quite as simple as that of the original formation of the tail itself, which no one disputes.

There is to be further considered the theoretical resistance of space to the projection and deflection of such enormous volumes of attenuated matter as appear in comets' tails. While it may not be absolutely necessary to offer an explanation of this apparent difficulty, in view of the fact that such projection and deflection do actually occur, still, the well-known laws of the diffusion of gases, in accordance with which any gaseous matter will traverse any other gaseous matter with the same velocity as, and with no more resistance than, in a vacuum, will show that this difficulty has been much overrated, while for the twin difficulty, how to account for the persistence of luminosity at such vast distances from its source, we may quote from Professor Proctor, " Cometic Mysteries," who, in turn, quotes as follows: "Comets travel in what must be regarded as to all intents and purposes a vacuum. From Dr. Crookes' experiments on very high vacua we may infer that there is very little loss of heat, except by radiation." By "intents and purposes" we understand, of course, as a cause of resistance, and certainly there is no reason to believe that the attenuated vapors of space are sufficient in density to cause any rapid diffusion of heat by convection, as contrasted with that of radiation.

We have seen that comets of short period sometimes disappear, and that their disappearance is frequently followed by the appearance of trains of meteors. In other words, they have apparently lost their cometic properties and become permanent adjuncts to our solar system. A curious confirmation of this fact is to be found in the character of the occluded gases which are contained in such meteorites as sometimes fall upon the earth's surface. Of this Professor Proctor says, "We have reason to believe that the nucleus of a comet consists of an aggregation of stones similar to meteorites." Speaking of the condition in which meteorites reach the earth, he says, "They are known to contain as much as six times their own volume of gases (taken at atmospheric pressure). In one of these meteorites recently examined by Dr. Flight, the following percentages of various gases were noted: Of carbonic oxide, 31.88; of carbonic acid gas, 0.12; of hydrogen, 45.79; of olefiant gas, 4.55; and of nitrogen, 17.66." The presence of olefiant

INTERPRETATION OF COMETIC PHENOMENA. 233

gas at once suggests the hydrocarbons of the cometic nucleus. The presence of this gas cannot be accounted for by the passage of the meteorite through our atmosphere, nor can that of hydrogen, and these are two characteristic gases, together with the vapor of carbon, constantly found to exist in comets.

As before explained, the advent of a comet into our solar system is that of a stranger, with electric polarity the opposite of that of the planetary electrospheres and identical with that of the sun. Under the combined influence of the solar gravity and perturbation by the gravity of the planets these foreign bodies tend to shorten their periods, and finally fall into the ordinary array of the bodies which compose our own solar system. But when this occurs they will, in turn, become contributors to, instead of antagonists of, the energy of the sun; in other words, they must then conform electrically to the condition of the family into which they have married,—that is to say, the planets,—and a reversal of their electrical polarity will take place. This reversal of polarity is no novelty in the operation of electrical apparatus. In "Electricity in the Service of Man" we read as follows of the Voss induction machine: "This machine is exceedingly powerful in favorable weather, but has an important defect *in a tendency to self-reversal, which is apt to occur at a stoppage.* This defect can be produced in a Voss machine, when desired, *by holding a metal point* to the positive brush K. The two derived inductive circuits are beautifully mani-

20*

fested when this machine is worked in the dark. A luminous stream is seen pouring towards the collecting comb L on whichever side of the machine the comb is positive." It will thus be seen that simple contact of a neutral (or negatively opposite) body will reverse the electrical polarity of this machine, or even the interruption of its motion will do so at times. Possibly a similar reversal may be produced in a comet by the contact in whole or in part of its nucleus with a planetary electrosphere, since the action of gravity is entirely independent of that of the attraction or repulsion of the electrospheres of both planetary and cometic bodies. Such reversal of polarity in a comet would at once extinguish its luminosity, and the generation of oxygen would at once replace the prior generation of hydrogen, and herein we may find explained the presence of carbonic oxide in large volume and carbonic acid in small volume in the meteorite above referred to, and of which gases Professor Proctor says, "It is quite certain these gases were not taken up by the meteorolite during its flight through the air." These aggregations of discrete meteoric bodies, loosely adherent by mutual gravity alone, would be gradually torn apart by planetary interference and dragged into streams of small bodies, thenceforth traversing space in elliptical orbits around the sun, just as do the planets and planetoids. Cyanogen, also, the deadly gas so frequently found to exist in enormous quantities in the nuclei of comets, would at once disappear, by double conversion into carbonic acid, or oxide, and ammonia, or nitrogen, so that this danger, as the result of a comet's possible approach to the earth's atmosphere, may be dismissed from apprehension.

It will be seen that all the enormous difficulties in the phenomena of comets find an explanation in the operation of the same universal laws which we have endeavored to apply to the other sidereal bodies. In conclusion, we may cite the following from Dr. Huggins: "Broadly, the different applications of principles of electricity which have been suggested group themselves about the common idea that great electrical disturbances are set up by the sun's action in connection with the vaporization of some of the matter of the nucleus, and that the tail is probably matter carried away, possibly in connection with electric discharges, under an electrical influence of repulsion exerted by the sun. This view necessitates the supposition that the sun is strongly electrified, either negatively or positively, and, further, that in the processes taking place in the comet, either of vaporization or of some other kind, the matter thrown out by the nucleus has become strongly electrified in the same way as the sun,—that is, negatively if the sun's electricity is negative, or positively if the sun's is positive. The enormous disturbances which the spectroscope shows to be always at work in the sun must be accompanied by electrical changes of equal magnitude, but we know nothing as to how far these are all, or the great majority of them, in one direction, so as to cause the sun to maintain permanently a high electrical state, whether positive or negative." The above speculations will have thus become demonstrated facts (though not in the mode suggested by the above writer) as soon as we clearly understand that, instead of the sun's "enormous disturbances" producing "electrical changes of equal magnitude," it is the electrical changes of equal magnitude which themselves cause the sun's disturbances, and that the sun's negative electrical polarity is permanently fixed by the opposite and positive polarity of the planetary electrospheres, and that all these various phenomena are but the normal expression of a single universal law, and are all due to the constant interaction of planetary, solar, and cometic electrospheres, in accordance with the well-established principles of electrical science. If, however, we consider, as is generally believed to be the case, the sun itself to be the sole prime source of its visible energy, nothing but difficulty and vague speculation can be looked for on every hand; but by relegating the solar orb to its proper place, and taking as the starting-point the true source of all energy,—to wit, the hidden forces embodied in the vapors or gases of interstellar space,—the whole process and mode of action will logically follow, and obscurity and difficulty together disappear. This principle, properly understood, is a master-key which will unlock every problem and interpret every enigma which the realms of interstellar space can present.

CHAPTER X.

THE RESOLVABLE NEBULÆ, STAR-CLUSTERS AND GALAXIES.

WHEN we come to consider the nebulæ, and endeavor to learn what part electricity has to play in the phenomena presented by these singular objects, we must recollect, in order to give them their due importance, that they are neither few in number nor uniform in constitution. Of the nebulæ, Professor Proctor (" Star-Clouds and Star-Mist") says, "When the depths of the heavens are explored with a powerful telescope a number of strange cloud-like objects are brought into view. It is startling to consider that if the eye of man suddenly acquired the light-gathering power of a large telescope, and if at the same time all the single stars disappeared, we should see on the celestial vault a display of the mysterious objects called nebulæ or star-clouds exceeding in number all the stars which can now be seen on the darkest night in winter. The whole sky would seem mottled with these singular objects." As telescopes, with the advances of constructive art, increased in power, these luminous clouds became more and more clearly defined, and many of them became resolved into clusters of stars, galaxies of suns like the Milky Way, of which latter our solar system is a constituent part, but more distant from us than the separately visible stars of that galaxy, and each separated from the relatively adjacent clusters by intervals of space comparable only with those which separate them from our own system. Of these glorious star-clusters, says Flammarion, in "The Wonders of the Heavens," "In the bosom of infinite space, the unfathomable depth of which we have tried to comprehend, float rich clusters of stars, each separated by immense intervals. We shall soon show that all the stars are suns like ours, shining with their own light, and foci of as many systems of worlds. Now, the stars are not scattered in all parts of space at hazard; they are grouped as the members of many families. If we compared the ocean of the heavens with the ocean of the earth, we should say that the isles which sprinkle this ocean do not rise separately in all parts of the sea, but that they are united here and there in archipelagoes more or less rich. . . . They are all collected in tribes, most of which count their members by millions." Says Professor Nichol, "System on system of majesty unspeakable float through the fathomless ocean of space. Our galaxy, with splendors that seem illimitable, is only a unit among unnumbered throngs; we can think of it, in comparison with creation, but as we were wont to think of one of its own stars." Of these glorious star-clusters the same writer says, "That no one has ever seen them in a telescope of adequate power without uttering a shout of wonder." These mist-like star-clouds were successively resolved, nebula by nebula, until science settled into the belief that with telescopes of adequate power all nebulæ might be so resolved, and the capacity of telescopes to thus resolve nebulæ became a test of their power. But spectrum analysis finally entered the lists with new methods of investigation, and the comparatively tiny spectroscope at a single leap passed far beyond the utmost limits of the highest telescopic vision, and at one blow struck the whole category of nebulæ into two widely different classes,—those composed of discrete stars grouped like the suns of our own Milky Way, and exhibiting the characteristic spectra of such bodies, and those composed of diffused gaseous matter not yet condensed into suns, and showing the disconnected spectral lines of simple elemental gases. The line of division was clear, direct, positive, and beyond all dispute. Yet beyond these two classes further research has disclosed certain vast nebulæ in which some portions exhibit true solar spectra more or less modified and others true gaseous spectra, each apparently merging into the other by gradations so faint and delicate that the inference is irresistible that in these nebulæ we see the processes of galactic and solar creation at various stages of their development.

Of these nebulæ, Professor Ball says, "In one of his most remarkable papers, Sir W. Herschel presents us with a summary of his observations on the nebulæ, arranged in such a manner as to suggest his theory of the gradual transmutation of nebulæ into stars. He first shows us that there are regions in the heavens where a faint diffused nebulosity is all that can be detected by the telescope. There are other nebulæ in which a nucleus can be just discerned, others again in which the nucleus is easily seen, and still others where the nucleus is a bright star-like point. The transition from an object of this kind to a nebulous star is very natural, while the nebulous stars pass into the ordinary stars by a few graduated stages. It is thus possible to enumerate a series of objects, beginning at one end with the most diffused nebulosity and ending at the other with an ordinary fixed star or group of stars. Each object in the series differs but slightly from the object just before it and just after it." And of these composite nebulæ, he adds, "The great nebula in Orion is known to be the most glorious body of its class that the heavens display. Seen through a powerful telescope, . . . the appearance of this grand 'light stain' is of indescribable glory. It is a vast volume of bluish gaseous material with hues of infinite softness and delicacy. Here it presents luminous tracts which glow with an exquisite blue light; there it graduates off until it is impossible to say where the nebula ceases and the black sky begins."

With reference to these distant galaxies of apparently complete solar systems like our own, the same principles must regulate the conversion of this energy of planetary electricity into the energy of solar light and heat as we see manifested in our own sun. The light of the individual stars is sufficient evidence of this; but the question may be asked, Is the electrical interaction between separate galaxies and between different solar systems in the same galaxy universal, or are these operations merely local? In other words, Is the source and the mode of solar energy in accordance with a single universal law of and between all created universes, or is it limited in effective energy to the members of each individual solar system alone? The answer is, that it is not less universal than the law of gravitation and no more so. There is a prevalent popular fallacy that the force of gravity is such that the movements, not only of solar systems, but of whole galaxies, and of all the illimitable systems of galaxies, are under its effective control, and that the whole universe of boundless space acknowledges its overwhelming sway. But nothing can be further from the truth. We know, of course, that the law is universal, as expressed in the statement of its terms by Newton, but the mere statement of the law itself, as applied to interstellar distances, refutes the idea that solar systems and galaxies can rotate around any common center by virtue of the attraction of gravitation as a controlling force. The universality of the law itself has even been doubted. Professor Ball says, "In the first book about astronomy which I read in my boyhood there was a glowing description. . . . I allude to the discovery, or the alleged discovery, of a certain 'central sun' about which it was believed or stated that all the bodies in the universe revolved. . . . It was too good to be true. No one ever hears anything about the central sun hypothesis nowadays. . . . It must be, then, admitted that when the law of gravitation is spoken of as being universal, we are using language infinitely more general than the facts absolutely warrant. At the present moment we only know that gravitation exists to a very small extent in a certain indefinite small portion of space. Our knowledge would have to be enormously increased before we could assert that gravitation was in operation throughout this very limited region; and even when we have proved this, we should only have made an infinitesimal advance to a proof that gravitation is absolutely universal."

Any one who chooses may prove for himself that the force exercised by gravitation between the multitudinous suns of our own galaxy, the Milky Way, and our earth must be quite infinitesimal, and totally unable to control the motions of our own solar system in a definite orbit through universal space. We know that the law which regulates the intensity of light at various distances is the same as the law of gravity,—that is to say, the proportion is directly as the mass and inversely as the square of the distance. We know also that the stars which compose the Milky Way are similarly constituted, generally considered, to our own sun, and that under similar circumstances the emission of light, roughly speaking, will vary according to the magnitude of these distant suns. Now, if any one will stand, at the darkest hour of the night, when the moon is absent and the sky perfectly cloudless, when the

> "Stars that oversprinkle all the heavens seem to twinkle
> With a crystalline delight,"

and sweep with his gaze all the concave hemisphere of the sky, and then compare the light which is radiated around him with the gorgeous effulgence of the noonday summer sun, he can pretty closely compare the relative attraction of gravity which all those distant suns together can exercise upon our earth with that of our own sun. Under control of the latter, the earth sweeps around in her orbit at the rate of about twenty miles per second; all these suns could not give our solar system even a minute fraction of that. Of this starlight Professor Ball says, "The sun certainly must receive some heat by the radiation from the stars; but this is quite infinitesimal in comparison with his own stupendous radiation." Any such attraction, of course, could not control the motions of our solar system, and much less that of many of the others.

> "The night has a thousand eyes, and the day but one,
> But the light of the whole world dies when the day is done."

We can also demonstrate the fact mathematically by an exceedingly rough calculation, which, however, will be sufficient for our purpose. Of the Milky Way, which comprises only the stars of our own sidereal system, Professor Ball says, "One hundred million stars are presumed to be disposed in a flat circular layer of such dimensions that a ray of light would require thirty thousand years to traverse one diameter." (The most recent estimates make the number of the stars which compose the Milky Way several times one hundred million, occupying a correspondingly greater amplitude of space. The number in any case is sufficiently stupendous.) Our solar system is located in space at the apex of a vast transverse cleft, and nearly at the center of this disk. Let us leave out of consideration the lower half of the Milky Way, as we look upward on a starlit night, and conceive this galaxy to extend only across the midnight sky above us like an archway, with fifty million suns, visible and invisible, exposed in the field of our vision. The nearest of all the fixed stars to us is that known as Alpha Centauri,—not visible, however, in our northern skies. This star is about two hundred and thirty thousand times as far from our sun as is the earth. If of the same mass as our sun, it must exert upon us an attractive force of gravity one fifty-three-billionth that of our own sun. Next in distance is the star No. 61 of the constellation Cygnus. This may be three times as distant, and is certainly not less than twice. The light of the former will reach the earth in three and one-quarter years; that of the latter in not less than six and one-half years, perhaps much more. These are our nearest stellar neighbors. While the former will attract us with only one fifty-three-thousand-millionth that of the sun, the latter will attract us with less than one two-hundred-thousand-millionth that of our sun. Conceive, then, a square pyramid extending radially upward for three thousand times the mean of these distances to the upper probable limits of the Milky Way, a light-distance of fifteen thousand years, and that this pyramid expands according to the squares of its distances, so that it will contain within it, equally distributed, all the stars (fifty million) of the upper half of the disk of the Milky Way; the sum total of all these attractions could not reach one twenty-millionth part of that of our sun upon the earth. If we continue to pile galaxies, in the same perpetual recession, behind each other to all infinity, we still could not engender sufficient attractive force to control the observed movements of the multitudinous stars of space. The very statement of the law of gravitation itself disproves it; for if we multiply orbs and systems according to any principle of aggregation that we know of in the way of distribution of such systems, or anything possible, with due regard to their own mutually interacting movements in space, we could never reach the inside limits of such a sphere of control, because the piling up of orb behind orb adds but an infinitesimal fraction to the force of gravity, for as the orbs themselves multiply in distance progressively by hundreds, their relative attractions inversely diminish by ten thousands. No possible increase of suns directly in mass could compensate for such an inverse ratio of squares, even if all intergalactic space were peopled with suns, instead of being, in fact, like a vast ocean, with a few small clusters of islands scattered here and there throughout its illimitable extent.

Of these vast realms of space, Professor Ball

asks, "Is our sidereal system to be regarded as an oceanic island in space, or is it in such connection with the systems in other parts of space as might lead us to infer that the various systems had a common character? The evidence seems to show that the stars in our system are probably not permanently associated together, but that in the course of time some stars enter our system and other stars leave it, in such manner as to suggest that the bodies visible to us are fairly typical of the general contents of the universe. The strongest evidence that can be presented on this subject is met with in the peculiar circumstances of one particular star. The star in question is known as No. 1830 of Groombridge's catalogue. It is a small star, not to be seen without the aid of a telescope. . . . We shall probably be quite correct in assuming that the distance is not less than two hundred billions of miles. . . . The velocity is no less than two hundred miles per second. . . . The star sweeps along through our system with this stupendous velocity. . . . The velocity being over twenty-five miles a second, the attraction can never overcome the velocity, so that the star seems destined to escape." Of the star Alcyone he says, "Doubtless that star is thousands of billions of miles from the earth; doubtless the light from it requires thousands of years—and some astronomers have said millions of years—to span the abyss which intervenes between our globe and those distant regions." And yet these stars, these galaxies, and even all the nebulæ we see or ever shall see, are merely in the vestibule of space; we have scarcely even yet lifted the outer curtain at the entrance of those vast realms. That the popular, but pseudo-scientific, idea of a series of ever-widening concentric orbits, increasing at every new expansion by an inconceivable ratio, is incredible we can well understand, and it is a satisfaction to know that such a wild hypothesis finds no warrant in the dicta or the demonstrations of science. And it is in the failure of gravity to control over the intervening space which lies between those vastly distant centers that we may hope to find the inklings of a more far-reaching law, by which nebulæ like that of Orion crystallize out into separate star systems, just as in the rocks, whether igneous, metamorphic, or sedimentary, we find the attraction of cohesion yield to that of crystallization, until the whole cleft rock blazes with countless garnets in the schist and quartz crystals in the gneiss, or reveals the yellow specks of olivine in volcanic ejections.

We shall find in the processes concerned with the development of living things the workings of a similar great law, perhaps the same. Wherever there is the possibility of life, there we find life. There seems to be an all-pervading vital tension, so to speak, an energizing force, which drives the evolution and ascent of life forward and upward by successive leaps, as it were, from type to type, from race to race, and even from nation to nation. In this universal forward movement we may dimly discern the primordial creative and developing impulse, constantly acting, but manifesting visible change only at intervals as gathering forces accumulate and equilibrium is disturbed. It manifests itself in all the fields of nature,—vital, chemical, molecular, molar, systemic. It is the ever-acting, eternal past, present, and future, the macrocosm and the microcosm, the panurgus, the Brahma, the Ancient of Days, and cannot be silenced or evaded:

> "They reckon ill who leave me out,
> When ME they fly I am the wings."

R. Kalley Miller, in his "Romance of Astronomy," says, "It would be hopeless to attempt expressing in ordinary language the vast distance at which these clusters of stars are situated from us. If we were to reckon it in miles, or even in millions of miles, figures would pile upon figures till in their number all definite idea of their value was lost. We must choose another unit to measure these infinitudes of space,—a unit compared with which the dimensions of our own solar system shrink into absolute nothingness. The velocity of light is such that it would flash fifteen times from pole to pole of our earth between two beats of the pendulum. It bridges the huge chasm that separates us from the sun in little more than eight minutes. But the light that shows us these faint star-clusters has been travelling with this frightful velocity for more than two million years since it left its distant source. We see them to-day in the fields of our telescopes, not as they are now, but as they were countless ages before the creation of man upon the earth. What they are now who can tell?"

The movements of solar systems through space are unquestionably controlled by some wider law than that of gravitation, and it still remains for science to seek its hidden principles and discover its mode of operation. We know that some stars travel alone, like the star already noted, No. 1830 of Groombridge's catalogue; that others travel in pairs, like the double star Mizar and its companion Alcor; and others in groups, like the stars Beta, Gamma, Delta, Epsilon and Zeta, of the constellation Ursa Major; that we are driving towards the constellation Lyra and leaving behind us Sirius and its fellows, and that many, if not all, of the stars whose motions we can measure have a rapid movement through space, but under what control, in accord with what hidden harmony, and under what general plan they move, we do not know; but the laws of electrical action of the circling planets upon their central suns, and of these upon space, we can readily account for by the similar operation of the same laws within our own solar domain; and we know by the similar terms of the ratio of distribution of light that this is commensurate in extent with the law of gravity, and operates in a like proportion of energy over all intervening distances; so that wherever our sun presents a visible point of light, there it is pouring its energy into space, and every sun we can see, every galaxy, every star-cluster, nay, every nebula, is likewise pouring into the interplanetary space of our own solar system its proportionate quota of energy. The very fact that we can see the star shine is itself the fullest evidence that this is so, and evidence also that the law of gravitation there, too, is still in force, operating over these same distances, and with the same proportionate energy.

Knowing all this, we can read with a new light the grand vistas of the skies, with their starry denizens, and claim them all as parts of our own family; and the mutual interchange of attractive energy and of light and heat will not fail between us until those inconceivable distances shall have been reached which human knowledge can never span and where speculation fails; and even there, from out those dark abysses,—dark to our human eyes,—the call will still faintly reach us, and our response will reach them also, though we shall never have tangible evidence that such mutual ties continue to exist. Industriously our planets gather their mighty energies from the surrounding springs of space, as one dips water from a crystal stream; we hand it over to our sun, and he, the royal high-priest, sprinkles it in glittering diamond-sprays over all those countless suns and their subject worlds, and they are baptized with an eternal baptism into our common brotherhood and we into theirs. Our familiar planets, Mars, Jupiter, Neptune, the earth, and even our little moon, seem to raise their voices and take actual part in the councils of almighty power, to move about as perpetual benefactors, gathering and spreading beneficence abroad, instead of cowering, a hapless few, like storm-stayed travellers, around the dying embers of our poor old sun, passive recipients of the light and heat and life which we have been taught to believe are slowly sinking into ashes and fading away in eternal darkness and death. One swift glance into these boundless truths is better for the human soul than the slow passage of whole hopeless centuries, which leave as their inevitable legacy on earth a vast and final catastrophe, in which everything that gave us light and heat and being must perish forever. Has it, indeed, come to this, that the last word which science has to offer is, "After us the deluge"? By no means. We have merely been endeavoring to measure the right hand of God by weighing and measuring a single isolated one of his countless multitude of suns.

It is as though one standing beside a great waterwheel should estimate its power and rotation by measuring the width and depth of the buckets and calculating the weight of water which its thirty-two receptacles contain, saying, "at its present rate in so many seconds it will cease to move." But we take him to the water-gate, and show it wide open; to the great dam above it which contains cubic miles of water; and still beyond that to the mighty fountains bursting forth with their rush and roar from the rock-ribbed fastnesses of the eternal hills, and pouring their unfailing flood-tide down forever and ever. And we do not pause even here: we show him the vapors rising from the spent water again, condensing into clouds, pouring down in torrents of rain among the hills, and that these continuously feed the sources of the fountains, which in turn supply the wheel almost to bursting. And so it is with the glorious mechanism of the heavens.

The source of solar energy is not to be found in the sun itself, but in his environment; and he himself, in all his glory, is but the king, crowned with gold, blazing with rich apparel, and scattering benefits among his satellites, not from his own private treasury, but who himself is enriched by the mighty tribute with which his willing subjects continually endow him, and to whom alone he owes all his pride and power and wealth and magnificence, and which he, in turn, so freely expends, transmuted in form alone, in the perpetual improvement and welfare of his domain. He is the faithful ruler, but not the creator; the beneficent monarch, but not the god.

CHAPTER XI.

THE GASEOUS NEBULÆ.

When we reach the irresolvable nebulæ, we unquestionably have approached the creative period of solar systems and in many cases of whole galaxies. These are multifarious in form, but all can be reduced to a few comprehensive types. In determining the question as to whether these irresolvable nebulæ were composed of distinct stars like the Milky Way, but too distant to be resolved from their mist-like light into discrete stars by the most powerful telescopes, or whether they were gaseous in constitution,—that is, composed of diffused gaseous elements not condensed into solar bodies,—the spectroscope became the final and infallible test. Of this instrument, thus used, Professor Proctor, in his "Star-Clouds and Star-Mist," says, "A very few words will explain the whole matter to readers who remember the three fundamental laws of this new mode of investigation,—viz., that, first, light from a burning solid or liquid source gives the rainbow-colored streak of light commonly known as the prismatic spectrum; secondly, when vapors surround such a source of light, the rainbow-colored streak is crossed by dark lines; and, thirdly, when the source of light is gas, there is no longer a rainbow-colored streak, but merely a finite number of bright lines." Dr. Huggins selected for investigation the small planetary nebula in the Dragon. He says, "When I had directed the telescope armed with the spectrum apparatus to this nebula, I at first suspected that some derangement of the instrument had taken place, for no spectrum was seen, but only a short line of light. I then found that the light of this nebula, unlike any other extra-terrestrial light which had yet been subjected by me to prismatic analysis, was of definite colors, and therefore could not form a spectrum. A great part of the light is monochromatic, and so remains concentrated in a bright line occupying a position in the spectrum corresponding to its color. Careful examination showed a narrower and much fainter line near the one first discovered. Beyond this point, about three times as far from the first line, was a third exceedingly faint line. From the position of one of the bright lines it is inferred the *gas nitrogen* is one of the constituents of the nebula; another line indicates the *existence of the gas hydrogen* in that far-off system; the third line has not yet been associated with any known terrestrial element, though it is near one belonging to the metal barium, and *still nearer one belonging to oxygen;* a fourth line occasionally seen *belongs to hydrogen.*" Professor Proctor says, "Dr. Huggins examined a large number of the planetary nebulæ (so called), obtaining in each case a spectrum which indicates gaseity. In some cases only one line could be seen, in others two, more commonly three, and in a few instances four. When these lines were seen they invariably corresponded in position with those already described. The single line sometimes seen corresponded with the brightest line of the three; and when a second line was visible, this also was no new line, but agreed with the second brightest line in the three-line spectrum. The fourth line was seen only in the spectrum of a very bright, small, blue planetary nebula, but was later observed in other cases, and especially in the great Orion nebula." At this time the latter was not visible, but when Dr. Huggins had opportunity to examine it, he says, "The telescopic observations of this nebula seem to show that it is suitable to a crucial test of the usually received opinion that the resolution of a nebula into bright stellar points is a certain indication that the nebula consists of discrete stars." Professor Proctor says, "A simple glance resolved the difficulty. The light from the brightest part of the nebula—the very part which under Lord Rosse's great reflector blazed with innumerable points of light—gave a spectrum identical in all respects with that which Huggins had obtained from the planetary nebulæ. Thus, what had been deemed boldness in Herschel—namely, that he should have associated the wildest and most fantastic nebula in the heavens with the circular and (in ordinary telescopes) almost uniformly luminous planetary nebulæ—was unexpectedly confirmed." The spectrum of this nebula has more recently been photographed by a long exposure in the camera of the prepared plate. Of the result,

Professor Proctor thus speaks, "The nebula is seen to be in great part gaseous, and, where gaseous, to shine in the main with the tints described above; but parts of the nebula are not gaseous, and those portions which are so are not all constituted in the same manner. . . . That portion which is called the fish's mouth gives a continuous spectrum; in other words, the same spectrum which we obtain from a star or a star-cluster. This is the spectrum arising from a glowing solid or liquid mass, or if from a gaseous body, then the gaseous body must be in a state of great compression. . . . But the stars thus forming must be immersed in the glowing gas forming the general substance of the nebula. . . . It would be absurd to suppose that the nebula is a flat surface; . . . nebulous matter lies also, in all probability (certainly one might fairly say), between us and the stellar aggregration as well as on the farther side." Further, the same author says, "If, as is probable, the luminosity of the gaseous portion of the Orion nebula is accompanied by but a relatively small proportion of heat, then the rays from the violet and ultra-violet part of the spectrum are likely to give us much more complete information respecting the constitution of these nebulous masses than can be derived from the visible part of the spectrum."

In the recent work of Professor Ball, "In the High Heavens," that author says, "There are, however, good grounds for believing that nebulæ really do undergo some changes, at least as regards brightness; but whether these changes are such as

Herschel's theory would seem to require is quite another question. Perhaps the best-authenticated instance is that of the variable nebula in the constellation of Taurus, discovered by Mr. Hind in 1852. At the time of its discovery this object was a small nebula about one minute in diameter, with a central condensation of light. D'Arrest, the distinguished astronomer of Copenhagen, found in 1861 that this nebula had vanished. On the 29th of December, 1861, the nebula was again seen in the powerful refractor at Pulkova, but on December 12, 1863, Mr. Hind failed to detect it with the telescope by which it had been originally discovered. . . In 1868, O. Struve, observing at Pulkova, detected another nebulous spot in the vicinity of the place of the missing object, but this also has now vanished. Struve, however, does not consider that the nebula of 1868 is distinct from Hind's nebula, but he says, 'What I see is certainly the variable nebula itself, only in altered brightness and spread over a larger space. Some traces of nebulosity are still to be seen exactly on the spot where Hind and D'Arrest placed the variable nebula. It is a remarkable circumstance that this nebula is in the vicinity of a variable star which changes somewhat irregularly from the ninth to the twelfth magnitude. At the time of the discovery in 1861 both the star and the nebula were brighter than they have since become.' . . . It must be admitted that the changes are such as would not be expected if Herschel's theory were universally true. Another remarkable occurrence in modern astronomy may be cited as having some bearing on the question as to the actual evidence for or against Herschel's theory. On November 24, 1876, Dr. Schmidt noticed a new star of the third magnitude in the constellation Cygnus. . . . The brilliancy gradually declined until, on the 13th of December, Mr. Hind found it to be of the sixth magnitude. The spectrum . . . exhibited several bright lines which indicated that the star differed from other stars by the possession of vast masses of glowing gaseous material. . . . September 2, 1877, it was then below the tenth magnitude and of a decidedly bluish tint. Viewed through the spectroscope, its light was almost completely monochromatic, and appeared to be indistinguishable from that which is often found to come from nebulæ. . . . It would seem certain that we have an instance before us in which a star has changed into a planetary nebula of small angular diameter. . . . Professor Pickering, however, has since found slight traces of a continuous spectrum, but the object has now become so extremely faint that such observations are very difficult. . . . For the nebular theory we require evidence of the conversion of nebulæ into stars." And not, it may be added, of stars into nebulæ.

Of the irregular nebulæ, Professor Proctor says, "It may well chance, as long since suggested by Professor Clark, of Cincinnati, and as more cautiously hinted by Dr. Huggins, that in the varieties of constitution observed in the irregular nebulæ, and the evidence such varieties afford of progressive changes, we may find not merely direct evidence of the development of suns and sun-systems from the great masses of nebulous matter, but even what would be a far more important and impressive result,—actual evidence of the development of so-called elements from substances really elementary, or, at any rate, one stage nearer the elementary condition than are our hydrogen, nitrogen, oxygen, carbon, and so forth. The peculiarity of the spectral indications of the presence of nitrogen and hydrogen in the nebula, that only one line of nitrogen and two or three lines of hydrogen are discernible, instead of a complete spectrum of either element as seen under any known conditions, seems suggestive of what may be called a more elemental condition of hydrogen and nitrogen." Whether this be so, or whether these peculiarities are due to self-obscuration, or mutual reversal of the familiar lines due to the enormous disturbances of the nebular mass which must exist, it is certain that there is one terrestrial substance, at least, which acts invariably, in combination and chemical affinity, as a simple element in inorganic chemistry, but which is, in fact, compound,—to wit, the hypothetical radical ammonium, which is closely allied with the simple alkaline metals potassium and sodium, forming with them a single group; and yet, while the others have always remained as fixed, primitive elements, the hypothetical element ammonium alone is a composite substance consisting of hydrogen and nitrogen, two of the invariable gaseous constituents of all these nebulæ. In comets we find, vaguely expressed, an occasional strongly marked sodium line, and also the spectrum of carbon; in these gaseous nebulæ we find, as yet, no trace of carbon, and this element is so closely allied to hydrogen in its chemical affinities and reactions as to suggest that it may be the same element or some alloy of it, or in some allotropic form, as we find to be the case with other simple elements under special conditions. In organic chemistry—the chemistry of organic life—we find almost innumerable compound radicals which act as simple elements in combination, but which we can combine and separate into their constituents at will; to all intents and purposes, in their various reactions, they behave as elemental substances, and were it not that our analyses are able to resolve them, as the spectroscope resolves the nebulæ, we might well believe that here also we were dealing with simple primary elements. It is almost certain that great discoveries in this field of chemistry are not far distant, which will recall with wondering surprise the now universally exploded fallacies of the "Philosopher's Stone" and the "Universal Solvent." Indeed, we may find in the electrical energies of the planets and the self-repulsive force of the electrospheres of the earth and moon possible grounds for investigating anew some of the abandoned tenets of astrology, in the hope that the light of science may disclose some basis, at least, for what, at one time,—and for nearly all time, in fact,—was the universally accepted belief, not only of the ignorant, but of those the wisest and most learned of their day and generation. If the planets by their position can cloud the sun, nearly a million miles in diameter, with spots, or shed the brilliance of the aurora borealis over all our skies, may they not also cloud the embryonic intellect, or charge it with energies for a career of prosperity or of disaster? May not the unseen currents, or the electric storms around us, or the vast electrical phenomena of the sun as well affect the sprouting germs of the husbandman or some abnormally rapid development of an insect pest as the light, the warmth, the moisture, or the cold, which, to our coarser vision, are alone apparent? Fancy and fallacy revel luxuriantly where science fails, but truth existed long before science was systematized, and the supercilious condemnation of once generally accepted views without examination is merely pseudo-science, and scarcely a single grade higher in the scale than ignorant superstition itself. And every new advance in knowledge requires a new overhauling of abandoned material, just as a new advance in metallurgical knowledge enables us sometimes to work over again our once-rejected mining dumps with decided profit. Indeed, science itself is but a collection of observed facts reduced to system, and among the shrewd and practical miners there is a well-known saying, "The ore is where you find it," which has frequently put scientific assertion to the blush.

A study of the beautiful mezzotint plates, from the drawings of the Earl of Rosse, contained in

Professor Nichol's splendid work, "The Architecture of the Heavens," will clearly disclose the forms, as revealed by a powerful telescope, of many of these gaseous nebulæ. Of such nebulæ, Appleton's Cyclopædia says, "Nebulæ proper, or those which have not been definitely resolved, are found in nearly every quarter of the firmament, though abounding especially near those regions which have fewest stars. Scarcely any are found near the Milky Way, and the great mass of them lie in the two opposite spaces farthest removed from this circle. Their forms are very various, and often undergo strange and unexpected changes as the power of the telescope with which they are viewed is increased, so as not to be recognizable in some cases as the same objects." An example of this is shown in Plate X. (Figs. 1 and 2) of Professor Nichol's work, which gives a greatly enlarged view of those shown in Figs. 1 and 2 of Plate IX. (For Fig. 2 of Nichol's Plate X., see illustration of nebula with double sun, in previous chapter.) Professor Nichol says, "In every instance examined, save one, the planetary nebulæ are nebulæ with hollow centers." The inference which this writer makes, that such a planetary nebula consists of "a grand annular cluster of stars," has been since disproved by the discoveries of the spectroscope, but the telescopic form remains true, and still awaits further interpretation. While the irresolvable nebulæ seem to seek some retired spot in space for their processes, like certain animals when incubating, this rule is not

THE GASEOUS NEBULÆ. 263

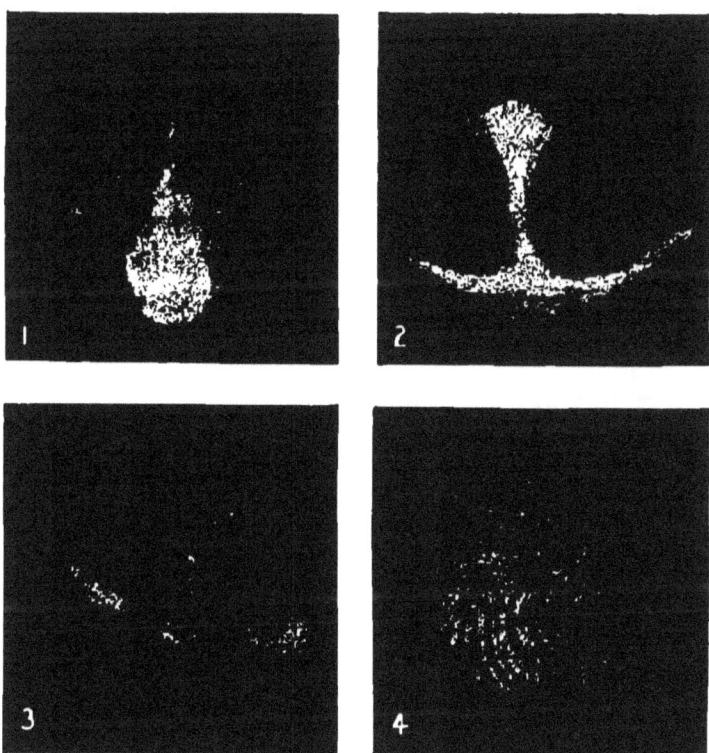

Gaseous nebulæ (non-systemic in development).—Fig. 1, the Crab nebula; Fig. 2, Dumb-bell nebula (reduced from Nichol, after Lord Rosse); Fig. 3, nebula in Sobieski's Crown; Fig. 4, Catherine-wheel nebula (from Flammarion).

In Fig. 1 gravity preponderates, and electrical repulsion drives the radiant matter upward and outward. This nebula resembles a comet in its phenomena; a large nebula in the neighborhood in rear of the Crab's body would produce this effect.

Fig. 2 shows a bipolar form produced by repulsion acting against gravity; the two heads connected by a narrow strand, the lower head elongated by internal repulsion, and the horns curved upward by the attraction of gravity of the upper head. This figure suggests the division of a comet (like Biela's) into two smaller comets.

In Fig. 3 gravity and electrical repulsion are nearly equal; the result is an elongated lineal nebula, warped into irregular curves by counter currents of space.

Fig. 4 is rotary, and the repulsive forces will probably entirely overcome gravity and result in the formation of an annular nebula with hollow center.

universal. Of this, Appleton's Cyclopædia says, "The density of nebular distribution increased with the distance from the galactic zone *for the irresolvable nebulæ*, but diminished with that distance for the clusters. . . . There is not a gradual condensation of nebulæ towards two opposite regions, near the poles of the galactic zone, but the nebulæ are gathered into streams, nodules, and irregular aggregations such as we find in the grouping of stars. . . . Between stars and nebulæ their arrangement follows the law of contrast. There are two remarkable exceptions to this law,— the Magellanic Clouds. In these, where stars of all orders, from the ninth magnitude to irresolvable stellar aggregations, are as richly gathered as in the galactic zone, nebulæ of all orders are also gathered richly, even more so than anywhere else over the whole heavens." In the same work, article "Nebula," it is stated of the planetary nebulæ, "There are several which have perfectly the appearance of a ring, and are called annular nebulæ. . . . Some appear to be physically connected in pairs like double stars. Most of the small nebulæ have the general appearance of a bright central nucleus enveloped in a nebulous veil. This nucleus is sometimes concentrated as a star and sometimes diffused. The enveloping veil is sometimes circular and sometimes elliptical, with every degree of eccentricity between a circle and a straight line. There are some which, with a general disposition to symmetry of form, have great branching arms or filaments with more or less precision of outline.

An example of this is Lord Rosse's Crab nebula. Another remarkable object is the nebula in Andromeda, which is visible with the naked eye, and is the only one which was discovered before the invention of the telescope. Simon Marius (1612) describes its appearance as that of a candle shining through horn. Besides the above, which have comparatively regular forms, there are others more diffused and devoid of symmetry of shape. A remarkable example is the great nebula in Orion, discovered by Huygens in 1656. .·. . The great nebula in Argo is another example of this class."

The number of nebulæ recognized in all the heavens is upward of five thousand, and new ones are being constantly discovered. Of these objects, Professor Nichol says, " The spiral figure is characteristic of an extensive class of galaxies. Majestic associations of orbs, arranged in this winding form, with branches issuing like a divergent geometric curve from a globular cluster." These nebulæ, however, are not associations of orbs; they are gaseous nebulæ apparently in process of evolution. This author (Professor Nichol) presents views of such spiral nebulæ either foreshortened to the view, so as to form a long ellipse, or with the convolutions of the spiral apparently raised from the horizontal plane into a conical form, and showing the black streaks of space which lie between the convolutions, others seen in side view, others in front, and, in fact, presented to the eye in every position for observation. The author wrote before the days of the spectroscope, and that he should conceive these vast objects to be spirals made up of blazing suns like our Milky Way—vast galaxies, in fact—was an inevitable conclusion at that time; but we now know that these spiral nebulæ are gaseous, are apparently in process of manufacture, and we can see them in their different stages of evolution, and may perhaps learn something about the processes by which solar systems and galaxies of suns are formed. Of one of these strange but exceedingly instructive objects, Professor Ball, in his work "In the High Heavens," says, "Fig. 3 represents one of the famous spiral nebulæ (that of Canes Venatici) discovered many years ago by the late Earl of Rosse. The object is invisible to the naked eye. It seems like a haze surrounding the stars, which the telescope discloses in considerable numbers, as shown in the picture. When viewed through an instrument of sufficient power, a marvellous spectacle is revealed. There are wisps and patches of glowing cloud-like material which shine not as our clouds do, by reflecting to us the sunlight. This celestial cloud is no doubt self-luminous; it is, in fact, composed of vapors so intensely heated that they glow with fervor. As I write, I have Lord Rosse's elaborate drawing of this nebula before me, and on the margin of this stupendous object the nebula fades away so tenderly that it is almost impossible to say where the luminosity terminates. Probably this nebula will in some remote age condense down into more solid substances. It contains, no doubt, enough material to make many globes as big as our earth. Before, however, it settles down into dark bodies like the earth, it will have to pass through stages in which its condensing materials will form bright sun-like bodies. It seems as if this process of condensation might almost be witnessed at the present time in some parts of the great object. There are also some very striking nebulæ which are often spoken of as *planetary*. They are literally balls of bluish-colored gas or vapor, apparently more dense than that which forms the nebula now under consideration. Such globes are doubtless undergoing condensation, and may be regarded as incipient worlds." Of these spiral nebulæ it is said, in Appleton's Cyclopædia, "Many of them had been long known as nebulæ, but their characteristic spiral form had never been suspected. They have the appearance of a maelstrom of stellar matter, and are among the most interesting objects in the heavens." Of their spectra it is said, " The bright-line spectrum is given by all the irregular nebulæ hitherto examined and by the planetary nebulæ." That is to say, these nebulæ are gaseous in constitution, and have not yet reached the stage of solar condensation which marks the existence of individual suns.

CHAPTER XII.

THE NEBULAR HYPOTHESIS: ITS BASIS AND ITS DIFFICULTIES.

"There sinks the nebulous star we call the Sun,
If that hypothesis of theirs be sound."—TENNYSON.

WHILE the nebular theory of Laplace is now the generally accepted scientific hypothesis of the formation of our solar system and of all solar systems, it finds its strongest support in the mode in which it seeks to account for the heat and light of the sun,—that is, that the central orb, gradually condensing down from an original volume as large as the orbit of Neptune, at least, after disengaging the planetary rings, continued to condense to its present volume, and still so continues, the molecular motions arrested by condensation under gravity reappearing in the form of the energy of light and heat, and that this process of degradation will continue until, finally, the sun becomes a solid inert mass, incapable by further condensation of exciting the ethereal undulations in space which constitute heat and light, when the whole process will finally cease, the sun will die out, the planets continue to rotate in darkness, and the whole machinery be left running through an eternal night, like a vast mill in the hands of a negligent watchman (or rather no watchman at all), left to run itself alone, dark, empty, lifeless, and deserted, through the long and silent watches of the night. While the source and mode of solar energy set forth in this work are to be as readily accounted for if we accept as valid Laplace's nebular hypothesis as by any other theory, yet such basis is not essential for its support; for while the planetary rotations and the central sun are the necessary consequence, according to Laplace's hypothesis, of their mode of formation,—are, in fact, just what we actually find them to be under any hypothesis,— electrical generation and transformation will proceed just the same whether the planets and sun were formed originally in one mode or in another. But, since this generally accepted hypothesis accounts for the light and heat of the sun, to a certain extent at least, and for a certain relatively brief period, while no other hypothesis has been able to sufficiently account for it at all, and while this hypothesis also finds both support and contradiction in many observed phenomena of our solar system, it may well occur that this hypothesis itself, based upon the necessity of accounting for the sun's light and heat, and which latter afford it its strongest basis of support, may, if the basis upon which the theory rests be found to be otherwise explicable, still remain as an end, while originally presented only as a means, and thus be held as an obstacle to the acceptance of the widely different interpretation of known facts herein presented, in the absence of any other hypothesis capable of explaining the same facts in accordance with this presentation of planetary electrical generation and the solar transformation of this energy into light and heat. Herbert Spencer mentions an instance of such perversion of means into an end as occurring during the agitation for the repeal of the corn laws in England, which extended over many years, during which organized efforts were made to influence Parliament. A permanent commission was established, with official head-quarters permanently located in London, with clerks, secretaries, higher officers, and all the paraphernalia of a first-class establishment. The purpose of this institution was to act in behalf of the popular interests upon Parliament by every available means to secure this great reform. After years of effort, he says, a clerk one day rushed, breathless, into the office from the House of Commons and shouted, in accents of despair, "We are ruined; the bill has passed!"

The nebular hypothesis, while generally accepted in lieu of a better one, has no actual primary basis beyond that of mere assumption. Of it Professor Ball says, "The nebular theory . . . seems, from the nature of the case, to be almost incapable of receiving any direct testimony." We have already quoted from Professor Newcomb that it must be accepted, with all its difficulties, until a different and sufficient explanation of solar energy shall be presented. As set forth in Appleton's Cyclopædia, the theory is as follows: "*Assuming, for the sake of the argument*, a rare, homogeneous, nebulous matter, widely diffused through space, the following successive changes will, on physical principles, take place in it: 1, mutual gravitation of its atoms; 2, atomic repulsion; 3, evolution of heat by overcoming this repulsion; 4, molecular combination at a certain stage of condensation; followed by, 5, sudden and great disengagement of heat; 6, lowering of temperature by radiation and consequent precipitation of binary atoms, aggregating into irregular flocculi and floating in the rarer medium, just as water when precipitated from air collects into clouds; 7, each flocculus will move towards the common center of gravity of all; but, being an irregular mass in a resisting medium, this motion will be out of the rectilinear,—that is to say, not directly towards the common center of gravity, but towards one or the other side of it,—and thus, 8, a spiral movement will ensue, which will be communicated to the rarer medium through which the flocculus is moving; and, 9, a preponderating momentum and rotation of the whole mass in some one direction, *converging* in spirals towards the common center of gravity. Certain subordinate actions are to be noticed also. Mutual attraction will tend to produce groups of flocculi concentrating around local centers of gravity and acquiring a subordinate vortical movement. These conclusions are shown to be in entire harmony with the observed phenomena. In this genetic process, when the precipitated matter is aggregating into flocculi, there will be found here and there detached portions, like shreds of cloud in a summer sky, which will not coalesce with the larger internal masses, but will slowly follow without overtaking them. These fragments will assume characteristics of motion strikingly correspondent to those of the comets, whose physical constitution and distribution are seen to be completely accordant with the hypothesis." During this process, it is further stated, successive rings of nebulous matter will be thrown off and left behind, which are supposed to have coalesced into planets and their satellites, and the motion of rotation will become more and more rapid as condensation proceeds, until, finally, the last planet, Mercury, will be left behind in annular form, and the sun will then become the central orb of all the planets, and condensation afterwards will proceed without further delivery of planetary rings. Professor Ball says, "If we go sufficiently far back, we seem to come to a time when the sun, in a more or less completely gaseous state, filled up the surrounding space out to the orbit of Mercury, or, earlier still, out to the orbit of the remotest planet."

There is nothing in the actively developing nebula illustrated on the following page which shows the slightest analogy, either in structure or the forces at work, to what is demanded by the nebular hypothesis. On the contrary, these radiating, spiral convolutions, springing from a center and extended, with interstratified dark spaces, out to the periphery, are entirely incompatible with that theory. There have not, so far, been observed in all the heavens any gaseous nebulæ which lend the slightest support to the nebular hypothesis. We should expect to find, if it were true, that many of the nucleated planetary nebulæ show exterior concentric rings of luminous matter, clearly defined,

Great spiral nebula in Canes Venatici. (See Fig. 156 of Guillemin's "The Heavens.") The small nebula to the right is also, according to M. Chacarnac, a spiral, though with the telescopic power used the figure above does not show it.

two, three, or a dozen in number, left behind by the contracting volume of the nebula, and coalescing into planets, and, within, the glowing disk from which new external rings are about to be left as a residuum. On the contrary, these nebulæ gradually fade away towards their margins, and imperceptibly disappear in the blackness of space. If they terminated abruptly, we might suppose that here, at least, was the orbit of a newly forming planet, but the regular and delicate gradation of luminosity from maximum to zero shows that no such sudden breaking off has occurred. In all these nebulæ we find every definitely marked structure to exhibit the operation of combined forces of gravity and internal repulsion nearly equally balanced, but each acting independently of the other. These phenomena are as universal as the forces of cohesion and repellent polarity in the "attraction particles" of cell-life which determine the segmentation, growth, and development of the living organism. We find here the primal modification and differentiation of material structure under the stress of directly opposite and interacting primitive forces, and it is doubtless the same whether in a cell or a system. It is not a residuum, but the *vis a tergo*.

It is well known that there are many and great difficulties involved in the nebular hypothesis. As for the genesis of comets, it will be at once seen that the theory will only account for such comets as never venture much beyond the orbit of Neptune, as well as those which have an orbital plane nearly coincident with that of the planets. But most comets come from illimitable space, far, far beyond Neptune's circle and at all angles to the plane of the planetary orbits; and we have already seen that a disk of space of the diameter of Neptune's orbit and half as thick (see Proctor's "Familiar Essays") would, to contain all the matter of our solar system equally distributed, have a density of only one four-hundred-thousandth that of hydrogen gas at atmospheric pressure,—that is to say, such a volume of the lightest substance we know of would make four hundred thousand solar systems like our own. This author inquires if such a mass could, under any circumstances, rotate as a whole, and adds, "Has it ever occurred, I often wonder, to those who glibly quote the nebular theory as originally propounded, to inquire how far some of the processes suggested by Laplace are in accordance with the now well-known laws of physics?" But the great primal difficulty is in the first assumption of the theory, which is not only entirely gratuitous, but physically impossible. It is that this great plasma of nebulous material—in the case of our own solar system not less than six thousand million miles in diameter—should have in some way become aggregated into a homogeneous mass of the requisite tenuity, complete and perfect, and ready for the succeeding stages of the process, in which, however, the law of gravity has hitherto had no active operation whatever; for, if gravitation existed and operated therein, such homogeneous mass could never have been formed, nor ever existed even if formed. The very forces which alone could have brought this vast mass together must have been the identical forces which afterwards broke it up into the sun and planets, and the operation of the same force must have prevented its original formation at all. According to the theory, it was like a horse-race, in which all the participants stood silent and motionless until the judge cried, "Go!" But the judge was the great creative force itself, and if the fiat reached to this extent, the same power could just as readily—nay, far more readily—have shot the sun and planets forth into rotation, as children scatter dough-balls, instead of holding in abeyance the control of universal law so as to (as a humorous writer speaks of the operations of a child in his investigation of a watch) "see the wheels go round." This is not nature's plan, so far as human knowledge goes. Of course these masses gathering to this great nebulous center, if acted upon by gravitation, would have at once condensed around the center as a nucleus, and if rotation ever commenced, it must have commenced then, millions of years, doubtless, before the outlying masses had even got within hailing distance. When masses of people assemble at a camp-meeting, the first comers take the best places, and the late arrivals have to circulate around in the woods; they do not all gather in a circle and then make a grand rush. That would be fair, perhaps, but it is not nature. And this, unquestionably, is how, if ever formed at all, these nebulæ must have been formed into systems.

The fact that the orbital planes of very many of these asteroids are greatly inclined to the common planetary plane, and still more greatly inclined to one another, points almost unerringly to the existence during their stage of formation of some powerful force either of internal repulsion or external attraction. That no sufficiently large body could have been present to exercise such attraction so far outside the general planetary plane is self-evident, and if there had been such source of attraction, while the orbital planes of the asteroids might have been deflected from the common plane, they could not have been forced apart so as to differ largely among themselves. Certainly nothing pertaining to the nebular hypothesis could have produced any such effects under any conceivable circumstances, and especially at so late a period of its progress, after all the principal planets had been completed. The only alternative is self-repulsion, and this could only have been due to the causes and their mode of operation already described in this work. In a modified degree these planes exhibit the same irregular orbital deflections as are so conspicuously visible in the orbits of comets, and they must have been unquestionably produced in the same manner. The barren bands or stripes in the area occupied by these asteroids, like the dark or vacant rings of the planet Saturn, may have been largely affected by the perturbing attraction of the neighboring planet Jupiter; but certainly no influence of that great planet (himself in the common planetary plane) could have operated to cast these forming planetoids into planes of diverse inclinations among themselves or to that of his own. On the contrary, his whole force must have been exerted to bring them into the closest harmony with his own orbital movements.

Omitting discussion of the technical difficulties in the application of the nebular theory to demonstrated facts, which may be found in the books, we may again repeat that this theory is not essential to account for the heat of the sun, which finds its real source elsewhere, while, nevertheless, the theory in itself is not incompatible with the views which we have endeavored to present and demonstrate. Certain phenomena, however, have been considered in prior quotations in this work which may aid us to roughly indicate the successive processes by which the evolution of solar systems and galaxies may be explained on another basis which requires no violent assumptions to be made and no suspension of any of nature's universal laws. The same operations which we see around us at the present time in our own system, if extended to the dimensions of a nebular aggregation, would probably present the same phenomena as those we find partially disclosed in the gaseous nebulæ, particularly the spiral, and these would naturally determine the final production of solar systems such as our own. The gaseous nebulæ, not spiral, and the mixed nebulæ also, would fall into their appropriate categories in the same general plan, and a consistent mode of formation would be presented from the beginning to the end of the different processes.

It should be observed that the spiral required by Laplace's nebular theory is essentially a *centripetal* spiral. The spiral nebulæ we see in the heavens, however, are *centrifugal* spirals. This is clearly shown in Plates XV., XII., and the frontispiece of Nichol's "Architecture of the Heavens," as well as in Plates XIII. and XIV. Plate XV.—the open spiral—is directly contradictory of any phenomena which could occur in accordance with the nebular theory of Laplace. The frontispiece shows the only form which such a nebula could assume at any stage of its career,—that is, a close spiral with

Spiral nebulæ, reduced from Nichol, after drawings of Lord Rosse. Fig. 1 is from Plate XV., Fig. 2 from Plate XII., and Fig. 3 from frontispiece of Nichol's "Architecture of the Heavens;" Fig. 4 is from same work, showing a similar development, from a spiral nebula, of a solar system with a double star for its central sun.

nearly circular convolutions. But while this particular form is not only in entire accordance with the hypothesis which we are about to suggest, being in fact one of the later and necessary stages in its progress, any such spiral as that shown in Plate XV. is utterly out of the question in the application of the nebular theory of Laplace or in any of the more recent modifications thereof.

The only hypothesis by which the various phenomena can be adequately explained must almost certainly be based upon the combined action of gravitation and electrospheric repulsion. We find in the corona of our own sun such phenomena manifested in the most striking degree, even in a completed system, and we can well understand that during the early stages of systemic development such phenomena would vastly transcend anything which we could now hope to observe around our own sun. We see this repulsion still more highly developed in the formation of the tails of comets. While these coronal rays are not visible to a distance of more, perhaps, than five million miles from the sun's disk, we have seen that the tail of Newton's comet was shot forth to a distance of ninety million miles in a few days, as it were in a moment, by the tremendous electrical repulsion of the solar electrosphere, and that this enormous tail, which, if composed of hydrogen gas alone (it was, of course, enormously more attenuated), would have contained a mass much more than equal to the weight of the sun, was swung around over an arc of one hundred and eighty degrees, giving a radial sweep of the tail over a distance of two hundred and eighty millions of miles in less than four days. And the tails of many other comets have largely transcended in dimensions that of Newton, above cited. We have learned much of the laws which regulate the development of storms, cyclones, whirlwinds, water-spouts, and other vortical phenomena in the atmosphere of our own earth, and can readily apply these principles to phenomena of vastly greater magnitude. We know that the matter of comets' tails is self-repulsive, as shown in multiple tails, as well as that it is repelled by an adjacent similarly electrified electrosphere,— that of the sun, for example,—as with pith-balls in the familiar class-room experiments; so that we can gather a very fair and complete idea of the processes of nature when dealing with such phenomena on a vastly more extended scale, in which our moments are measured by millions of years and our miles by the almost infinite distances of sidereal and nebular space.

CHAPTER XIII.

THE GENESIS OF SOLAR SYSTEMS AND GALAXIES.

THE processes of development of a solar system from the diffused elemental matter of space may then be roughly sketched as follows, premising that each stage may have possibly extended over vast periods of time, and the whole, perhaps, not been completed for millions of years. With the processes of creation time is as nothing.

The area of space in which a solar system is about to be developed has hitherto maintained its molecular constituents in a state of gradually increased unstable equilibrium, whether such augmented instability may have been induced by a gradual rise of temperature from emission of the solar energy of other galaxies, by gradual diffusion from constantly operative centers, from currents or vortices of space, or by some primal inherent constitution of space itself, with constantly increasing tensions relieved by successive discharges, of which analogous instances are found in various other processes of nature, as, for example, ovulation, fission, and gemmation in the reproduction of life, regularly recurring epileptiform convulsions, regularly repeated spark discharges from electrical machines, or the ebullition of viscous fluids with their slowly recurring bursting bubbles. At some focal point of this area a rupture of tension will finally occur, induced by some sudden current or vortical movement, as we see sometimes in a pool of water gradually reduced in temperature below the freezing-point, when its whole surface, by the passage of a breath of wind even, will be suddenly flashed into crystals of ice. At this point of space there will be instituted a rapid expansion among the molecules and a consequent fall of temperature, followed by an inrush of the vaporous material surrounding this center of agitation, and a vortical movement will be established, with currents of spatial matter attracted to this vortex in constantly increasing streams. The molecular tensions will be successively unlocked as the circles of agitation continue to widen, and a condensed nucleus will form, rotating upon its axis and exhibiting the combined phenomena of gravity and centrifugal force. As the nucleus continues to increase in mass and density its temperature will constantly rise, while its speed of rotation will gradually diminish as its volume increases, and the aqueous vapors of space, as they gather around this rotating center of attraction, will be forced outward by centrifugal action and the heat of the nucleus, and form vast attenuated clouds,—not necessarily visible, however, to human sight,—and these clouds, in their various stratifications and disturbances, will gradually come to partake of the rotatory movement of the center, such movements, however, gradually fading away as they recede in space and in density. The cyclonic movements of these clouds of aqueous vapor upon themselves, but principally against the surrounding gases of space still under tension, will generate enormous quantities of electricity, which flash like thunder-clouds as they approach each other, with incessant streams of lightning and rolls of thunder. The growing and heating central nucleus is thus thrown into a state of high electrical opposite polarity, and its own constituent elements become self-repellent, just as we see in the sun's corona and in the phenomena of comets. The electrical tension of the central mass will gradually grow higher and higher, until a vast stream or streams of incandescent nebulous matter (for with double suns they may be multiple, or the internal repulsion may even cause division of the nucleus itself) will be suddenly driven outward in a radial direction along the lines of least resistance,—that is to say, in the plane of equatorial rotation, where centrifugal force is most effective. We can readily understand the self-repellent force of such an enormous mass of cosmical matter by considering that, in our own completed system, the repulsion of the solar electrosphere drove forth the tail of Newton's comet, as before stated, to a distance of ninety million miles, and whirled it around a semicircle of this radius in less than four days. Our most distant planet, Neptune, is only thirty times this distance from the sun, and we see during every solar eclipse the coronal structure glowing to a distance of more than a million miles from the sun's disk, and the radial streamers driven forth five million miles, and even farther. (See illustrations of solar corona in Guillemin's "The Heavens.") The vast stream of radiating nebulous matter thus forced out by solar repulsion will likewise be acted upon with equal energy by its own internal self-repellent force. If we conceive a stream of water thrown vertically upward by a powerful force-pump, in which every drop of the fluid is endowed with tremendous self-repulsive energy, we should find an analogy to the phenomenon in question. We can see an example of this in the "Crab Nebula," illustrated in a previous chapter. The stream, acted upon by gravity downward, by the force of ejection upward, and by the internal force of repulsion both transversely and upward, would assume a pyriform shape, narrower beneath, largely swollen about its middle, and thence gradually decreasing in diameter to its termination in a rounded tuft, in advance of which would be driven forth detached sprays and wisps, while filaments and outlying parallel strands would mark its entire ascent, except towards its point of ejection, where the primal force which drove it out is greatly in excess of those of gravity and self-repulsion. It will be seen at a glance that these phenomena are precisely those which we observe in a comet's tail. (See illustrations of many comets having these characteristics in Guillemin's "The Heavens," Lockyer's edition.)

Suppose, now, that this stream of water or the tail of a large comet were gradually wrapped around its point of emission by the rotation of this nucleus upon its axis. A spiral would form, very open or flaring at first, but gradually growing closer and more circular as the force of gravity drew its convolutions downward upon the interstratified clouds of aqueous vapor occupying, in compressed layers, the spaces between the adjacent coils of the spiral. There would be a composite action of forces observed: gravity would attract the convolutions and their interstratified layers of cloud equally, according to their densities, while the central repulsive force would repel the convolutions of the spiral along the same lines of force, but would not act at all upon the strata of clouds, and the force of internal self-repulsion would also tend to disrupt the convolutions of the spiral by expanding them outwardly. The outer convolution, however, would have no backward thrust from any internal repulsion beyond, while, within, gravity and solar repulsion would be more equally balanced, so that the outer coil would be relatively compressed in its rotation against the next inner convolution, and its ratio of distance would not be maintained. We find this exemplified in the case of Neptune's orbit in our own system. The inner convolution would also be abnormal, since the primal force of ejection must have been sufficient to carry the outward thrust of the whole spiral, and in consequence its flare would offer much greater resistance to the deflection of rotation, and it would have a more radial direction than those beyond. We shall find that the planet Mercury, and the inner convolution which was eventually reabsorbed into the solar mass, exhibit these phenomena. Between the outer and these inner convolutions the curve of the spiral would be approximately regular, with a fixed ratio of increase. In the planets of our solar system this ratio is that produced by constantly doubling the preceding number, the series being 0, 3, 6, 12, 24, etc. In other solar systems, however, the ratio may be quite different. In this abnormal flare of the inner convolution is doubtless to be found the rational basis of Bode's empirical law .of planetary distances, in which the arbitrary number 4 must be added to each term of the above progression, making the series 4, 7, 10, 16, 28, etc. The inner coil between Mercury and the sun was drawn into the solar mass on the disruption of the spiral, leaving, from the abnormally radial curvature of the inner portions of the spiral and its absence from the series, a vacant place which must be represented by the relatively fixed increment to be added to each term of the series.

As the convolutions of the spiral become more and more compressed towards each other and more and more flattened against the interstratified cloud-layers, the force of internal repulsion becomes more and more active in its tendency to disrupt the spiral, since its forces are more direct and concentrated along lines nearly at right angles to the force of gravity. During the formation of the spiral we can easily conceive that—like a stream of water shooting over a cascade, or the multiple tails of some comets, or even a whole comet, as, for example, Biela's, which was split up into two separate bodies by this force—some convolution, perhaps a single one of the series, will be laterally divided into a large number of nearly parallel strands, mutually held apart by their internal self-repulsion, and with cloud-layers interposed between these lateral strands. Such a series of small planets as

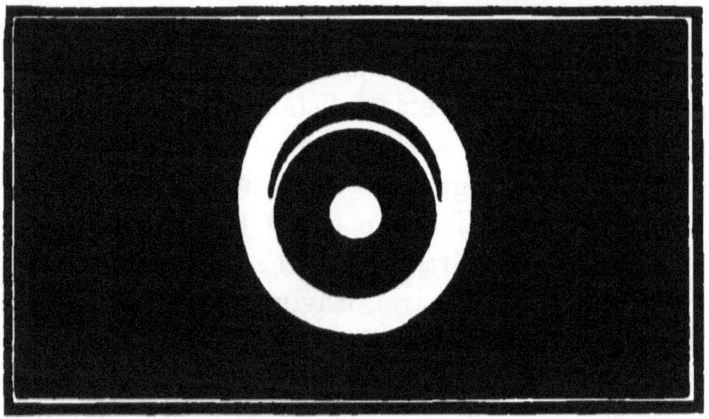

Nebula in Canes Venatici, showing central nucleus and external ring split and held apart by electrical self-repulsion. (From Helmholtz's "Popular Lectures.")

these would finally produce we find in the belt of our asteroids, the bulk of the convolution, probably, for the most part, however, scattered in space, since their aggregate mass is so small, and possibly, in part, coalesced into the mass of Jupiter, to which Mars, by his position, may also have contributed.

Not only may a whole convolution be thus split up, but along the spiral at many points the outer margins may be thrust outward, forming partially detached parallel strands, which may thus coalesce to form the satellites of the completed planets; while at the outer extremity of all, where the backward thrust of self-repulsion is wanting, enormous wisps, sprays, and tufts of nebulous matter would be driven entirely forth into the illimitable realms of outer space, but not necessarily, or even probbly, into the space of other systems, which are so enormously distant; and there, in those unoccupied realms, they will remain to gyrate "in the solitude of their own originality," in the form of comets, until, at long intervals, they may chance to revisit the scenes of their earliest youth, to warm their frozen limbs for a brief period at the old and well-remembered parental fire, or finally, worn out with toil and travel, "come home at last to die."

Driven forth from the society of their fellows by their own unbalanced energies, these anarchists of the sky may form loose aggregations, granulated about multitudes of self-constituted minor centers; but, cut loose from all effective solar control during their period of coalescence, they must forever lack the consolidated form and complex organization of their prosperous and rotund brethren, the planets and their satellites, or even the tiny asteroids, who stayed home and, like the little pig, had bread and butter for breakfast.

The disruptive energy of internal repulsion, as above stated, increases in force as the convolutions of the spiral become more and more compressed and the spiral becomes more and more circular in form. Suddenly the coils of the spiral will be burst asunder, and this will occur along that particular radial line of gravitation where the central nucleus acts with its most effective force. The disruption will be simultaneous, as a general rule, in accordance with the principles which control ruptures of tension of bodies in a state of unstable equilibrium, and which we see exemplified in multiplied centers of crystallization, the simultaneous formation of mud-cracks, the Giant's Causeway, and other like phenomena. Each convolution will now become a detached open ring, one of its broken extremities, however, being millions of miles farther from the central nucleus than the other. What occurs when a cometic body, negatively electrified, impinges upon the positive electrosphere of a planet, or when an electrical induction machine like Voss's is touched by an oppositely electrified body, will now necessarily occur with these disrupted convolutions. Their connection with the negatively electrified nucleus being broken, a reversal of electrical polarity will ensue from contact with the adjacent positively electrified clouds of aqueous vapor, and, instead of self-repulsion, mutual attraction will now prevail along the length of each of the open rings. Held apart from the central nucleus by the interstratified cloud-layers, and acted upon by the double force of gravity and internal attraction, the component elements of these open rings will rapidly lose their luminosity and heat, and coalesce by a retrograde movement down the lines of their direction, thus approaching the sun along the segment of an ellipse, the nucleus, or sun, occupying one of the foci, the eccentricity of the ellipse being measured by the differential between the nearest point of the open ring and the part of the convolution which lies directly opposite and beyond the sun. In other words, the form of the spiral will determine the eccentricity of the ellipse, subject to perturbations, however, of various sorts. During this stage of coalescence from an open ring into a sphere, these bodies will take on, by cooling and condensation, their planetary forms; and as the forming spheres, by the retreat of their masses down the lines of approach to the sun, advance, their forward and nearer extremities will be more powerfully acted upon by gravity than those parts in the rear, and a forward plunge or axial movement of rotation will be set up. Viscous matter,—pitch, for example,—molten by the sun's heat and flowing down a steep roof, exhibits a similar forward movement, the outer layers tending to roll over the inner ones in convoluted folds, the adhesion to the roof of the under surface corresponding to the retarding pull of the sun's attraction. In like manner are produced rotating eddies in streams of water having crooked channels, eddies of air under water-falls, and other analogous atmospheric disturbances. During the stage of coalescence of the planetary spheres the adjacent clouds of aqueous vapor will condense around them, and their hitherto diffused electrical energies will be concentrated by rotation in curents of enormous quantity and potential directly upon the sun, and a disassociation of the elements which compose these watery vapors will ensue, the result of which will be the deposit of hydrogen gas as an atmospheric envelope around the sun's body, and of oxygen around and through the bodies which constitute the planets. These gases will be disassociated in their combining proportions, two volumes of hydrogen at the sun for one volume of oxygen, distributed according to their relative electrical energies among the planets. This nascent oxygen will rapidly combine with the consolidating elements of the planets and, interpenetrating their solidifying bodies, form the vast mass of oxides which we find to constitute the bulk of our terrestrial mass, the residue, mechanically commingled with the condensed ever-present nitrogen, forming the planetary atmospheres. The condensation of volume of the planets will give rise to great elevation of temperature, while their currents of electricity, poured into the sun, will, by their passage through its enormously compressed hydrogen atmosphere, produce intense heat, and this, rapidly communicated to the solar core within, will raise its temperature to that of the sun as we now see it, and permanently maintain it in that state of incandescence.

During the stage of coalescence of the planetary bodies, outlying strands of the spiral will follow the course of their adjacent masses in a nearly parallel movement, and will gradually coalesce into smaller bodies more directly under the influence of the gravity of their own adjacent planets, by their proximity, than of that of the sun. These bodies will thus rotate as satellites around their planets, and the forward shift of their centers of gravity, by their advance along their lines of coalescence, may result in a permanent displacement, of which we see an example in the moon, which constantly presents the same face to the earth, while having an axial rotation of its own with reference to the sun. (In this case the action of gravity may have been assisted, however, by the mutual repulsion of the lunar and terrestrial electrospheres forcing the atmosphere and moisture of the lunar mass to its opposite side and maintaining it there, where it would remain as a buffer against rotation.) In some cases we might find certain outlying strands of a convolution which, perturbed by external influences, may have been delayed in its conversion into spherical form, and this subordinate strand, pyriform itself, as it must have been, in shape, would thus form a spiral of minute discrete bodies, probably like the nucleus of a comet, finally assuming the shape of a series of rings, and rotating like a satellite around the neighboring planet, the inner and outer strands more attenuated and the middle ones more condensed, as we find to be the case with the rings of Saturn.

In the original spiral we have seen that, as a whole, it was of necessity pyriform in shape. The planets formed therefrom would thus be found to increase in size from within outward to a maximum, after which they would again decrease, but not to the original minimum, while the extreme outer planet would also be unduly enlarged by increment from partially dissipated terminal filaments, gradually attracted thereto from surrounding space. There is such an undue enlargement of the planet Neptune, and this, with its relatively compressed orbit, before alluded to, renders it almost certain that Neptune is in reality the outermost member of our planetary system. We find this gradation of size to be the case in our solar system, except where the series has been broken by the multitudinous separation, from violent internal repulsion, of one of the convolutions into parallel strands showing all sorts of perturbations, this being the convolution which occupied the region between the orbits of Mars and Jupiter, and which, by the coalescence of these numerous parallel strands into small planetary bodies, has filled the space with a belt of asteroids hundreds and perhaps thousands or even tens of thousands in number. It is probable that a law regulating the ellipticity of planetary orbits can be deduced from a consideration of the principles which have governed their inception, and with these are doubtless closely related those laws of Laplace which have demonstrated that " in any system of bodies travelling in one direction around a central attracting orb, the eccentricities and inclinations, if small at any one time, would always continue inconsiderable." (Appleton's Cyclopædia, article "Planet.")

We have thus traced the genesis of a solar system from its earliest stages forward through its various changes until, complete and in working order, it is ready to be sent on its eternal course, either alone or as one of a vast congeries of similar systems, like the Milky Way. (See frontispiece for illustration of a series of types of development from a straight-tailed comet, through different curvatures, and spiral nebulæ of less and less divergence, until nearly circular, and finally terminating in a complete solar system.) These processes of creation may be isolated, or they may flash a hundred million solar systems into being together, as crystals flash forth in the rock; but, when once formed, they go forth each as eternal as space itself.

But can we not go back one step farther still in the progressive stages of creative energy? Whence came these powerful agencies by means of which all those distant regions became peopled with suns and worlds? The great source of all is to be found alone in space,—the so-called "empty space." But it is far from empty; all through it are diffused the attenuated vapors which, condensed, constitute our suns and planets, and all that is, or ever shall be, gaseous vapors, which are held poised, with their opposite tensions of cohesion and expansion, like the Prince Rupert drops which glass-blowers make for toys,—a little bulb of glass, chilled as it falls, molten, in a vessel of water. From one extremity projects a long, crooked stem, scarcely thicker at the end than a horse-hair, spun out from the molten glass as it hung from the glass-blower's rod. The bulbous body is as large, perhaps, as a nut; you can beat it with a hammer and it will not break; it is the hardest in structure of all glass. Now, wrap this bulb up in a thick handkerchief, or you may be injured; hold it firmly, and break off the very tiniest tip of the long stem three, four, or even six inches from the bulb. There is a sudden shock; open your handkerchief, and lo! instead of the solid bulb, there is only a loose mass of white powder. If you put the bulb in a heavy glass vessel full of water and break off the tip of the tail, it will shatter the vessel into fragments. What is the explanation?—it is, of course, well known—simply that the molecules of glass were instantly arrested in their motion of adjustment as the glass was suddenly chilled by the water, and the molecular motion of shrinkage was arrested, leaving the individual molecules under a tremendous strain of position in their endeavor to reach their true places. They are rigidly fixed in this position of unstable equilibrium, one balancing the other; but let a single molecule be displaced,—a fragment so tiny that the eye can scarcely see it,—and the molecules, thus thrown out of mutual support against each other, must now rearrange themselves from the ruptured rigid mass, and, like a row of stood-up bricks, each of which thrusts the other forward, with a sudden explosive force the molecules assume their true position of stable equilibrium, but it is at the cost of the whole structure. To this same cause we owe the explosive force of our gunpowder, nitroglycerin, and all explosives; the molecules are held in unstable equilibrium, and the tension once relieved at a single point, be it ever so infinitesimal, the molecules of the whole mass rearrange themselves with explosive energy. Strange that so harmless a substance as glycerin, by the mere replacement of an atom of nitrogen gas, should develop the energy of dynamite under a trifling molecular shock.

So, also, the aqueous and perhaps other vapors of all space, attenuated though they be, and perhaps by reason of this very tenuity itself, as shown by the experiments of Professor Crookes with attenuated gases when acted upon by electricity, are held in the same state of unstable equilibrium. We know the potency of this instability from the terrific explosive combination of the gases which combine to form aqueous vapor. We may again refer to one of the well-known experiments of Professor Crookes with simple atmospheric air. Enclosed in a cylindrical glass vessel, the electric spark passed freely; as it became more rarefied under an air-pump, new phenomena appeared, until, at a stage of high rarefaction, the molecules of these gases were driven forward by the electric current with such energy as first to raise the temperature of the opposite side of the cylinder to a red heat, then to melt, and finally to perforate the glass. The explanation is that the movements of closely aggregated molecules mutually interfere with each other; as they gain elbow-room by being reduced in number, they act with more directness, and consequently with more force: it is the difference between men fighting in a crowded room and out in an open field. It is possible that these molecular tensions of space, by the ready unlocking of the forces with which they are charged, may even aid in the rotation of the planets by acting upon their electrospheres in their drift through space, as charged thunder-clouds react upon each other, or the molecules of atmospheric air, in moderately high vacua, under electrical excitement, act upon the walls of the containing vessel, as in the experiments of Professor Crookes and others. The riddles of nature are like those of the sphinx,—they have more than one meaning.

The tensions of the aggregated molecules of space are thus counterbalanced only so long as all space is equally occupied and a state of perfect quiescence exists in its every part. A molecular disturbance in one part is immediately communicated to adjacent parts, and finally to all. With the first movement, gravity asserts itself, for gravity exists and must exist in all parts, and must actively manifest itself whenever the perfect mutual balance of space is disturbed and a center of energy developed, and co-ordinately with the action of gravity begins that of electricity. Movements among the molecules are converted into movement of mass; centripetal motion begets condensation, this begets sensible heat and vortical movement; then come the phenomena of electrical generation by moving contact with the gases of space, then repulsion and disassociation of the elements of the aqueous vapors, combination of simple into compound elements; and, the balance once disturbed, the state of unstable equilibrium is forever destroyed, and all space henceforth must exhibit constant change. There are whole segments of space absolutely blank, so far as visible systems are concerned, which seem to have been exhausted, for the present æons at least, to supply material for the vast adjacent galaxies which extend along their borders; see illustrations in Proctor's "Essays on Astronomy," article "Distribution of the Nebulæ."

It need not be supposed that such stage of perfect and universal quiescence ever existed in fact; it is like the Nirvana of the Buddhist philosophers, —a subjective and not an objective condition. We can have no knowledge of the existence, even, of material things, save from their phenomena, the manifestation of interchanging forces, upon which rests our threefold basis of knowledge, perception, cognition, and comparison. We know nothing of matter, except as affected by internal or external force, nor of force itself, except as it acts in one mode or another upon matter. All beyond this is, for us, without form and void.

Progressive change has always, doubtless, been the universal law of creation, and the great ocean of space is, and ever has been, and ever will be the highway through which perpetually plough the great caravels which bear the fortunes of creative energy, laden with life and light and heat, in their eternal progression. The creative impulse once given, if it, too, was not primeval in the eternal past, must have gone on from development to development, like the transmission of life, from age to age and from realm to realm. "The mills of the gods grind slowly;" in these vast areas time is absolutely nothing; the processes we see are but as the dip of a swallow's wing compared with an inconceivable futurity; but all our energies, and all the energies of planets and suns and systems and galaxies, and of whatever other and wider created forms may stretch onward to infinity, came forth from the ocean of space, and to this ocean all these energies continue to return again in ceaseless circuit.

Can we indicate any relationship of periodicity for the genesis of solar systems from space? There is a remarkable example of a somewhat similar periodicity in organic life for the rupture of tensions, so common that its analogous character and perfect regularity are scarcely even thought of. Among the highest species of mammalia we find that, in a state of health, whether resident of the heights of the Andes, the deserts of Africa, the jungles of India, or the most densely populated centers of London; among rich or poor, high or low, idle or industrious, virtuous or vicious, ancient or modern, civilized or barbarous, black, white, red, or yellow, the ovum of the mature female rises to the surface of the ovary, and at intervals, almost uniform, of twenty-eight days, organic excitement ensues, the enclosing vesicle is ruptured, and the ovum escapes. The remarkable feature is not that these processes continuously succeed each other; but that under such diverse conditions and opposite circumstances, and with two separate ovaries operating at the same time, simultaneously

THE GENESIS OF SOLAR SYSTEMS. 301

or successively, this almost miraculous interval of no more and no less than twenty-eight days between the successive ruptures of tension and their attendant phenomena, should constantly persist. For its ultimate cause we must look back to the *vis a tergo* to which we have already alluded; and there may be, and doubtless is, a similarly acting remote cause which regulates the periodical development of solar systems or of galaxies, periods of intense activity, followed by intervals of exhaustion and recuperation, and again succeeded by another period of activity, and so on perpetually, for space is perpetual, infinite, and inexhaustible.

It will be observed that the processes above roughly sketched are somewhat similar to those observed in the formation of so-called water-spouts, which usually terminate in dissipation in the atmosphere, or else in terrific thunder-storms, but which occasionally reach a sufficient energy of rotation to spin their central nuclei down towards, or even to, the surface of the sea, or, in desert regions, to that of the ground. There is no analogy with the theoretical and "assumed" primal mass of attenuated plasma of the nebular theory, or with its slow initial rotation, with the successive casting off of rings of nebulous matter. It may sometimes happen, however, that the repulsive electrical energy of the central nucleus may throw off its external envelopes with sufficient force to drive them entirely beyond the effective limit of its attractive forces, as occurs in the formation of embryonic comets as above described; in such case the nebula will be a variable one, with successively repeated aggregations and successive outbursts, periodical like the active stages of volcanoes; and, even when the nucleus has already presented a continuous solar spectrum, its energies may be thus expended, or more gradually, and finally dissipated like the electricity of a highly charged Leyden jar exposed to a moist atmosphere.

As a bottle of strongly effervescing liquid may blow itself empty, when suddenly opened, by the mutually repellent energy of its contained molecules, so if such a phenomenon were manifested in a radial direction from a central point, the repelled spray would show itself as a nebulous ring with a hollow center. An example of this sort is shown in the multiple-tailed "Catherine-wheel" nebula (Fig. 4 of a previous illustration). If such an annular nebula should become ruptured into two portions by internal repulsion, the electrical polarity of the smaller fragment would be reversed, and the two arcs would separately coalesce and consolidate into a sun and a single planet, forming a solar system like that of Algol, which has been already described. Otherwise, the nebula would probably retrograde and disappear, by diffusion, into space again. We may expect to find abortive efforts of nature here, as we so constantly find them elsewhere, not merely in inorganic matter, but even among the processes of life.

In Professor Proctor's article ("Essays on Astronomy") on the square-shouldered aspect of Saturn, he mentions a hitherto unexplained circumstance of the earth's atmosphere—the curious fact that the barometrical pressure of the earth's atmosphere is somewhat higher between the poles and the equator than immediately over the latter, as might be supposed to be the case. This is a phenomenon of mutual repulsion similar to those manifested in the operations above described. The rotation of the earth on its axis forces the terrestrial atmosphere, by its centrifugal motion, in undue proportion, around the equatorial belt, causing the same sort of atmospheric thinning at the poles which we see in the solar photosphere at its corresponding parts. At the same time the highly electrified atmosphere, by its mutually repellent action, tends to force this swollen equatorial ring backward toward the poles. The resultant of these two repulsions is an area of maximum density part way between the poles and the equator. It is probable that this self-repellent equatorial swell may play some part in the sun's atmosphere, in extending, and also in limiting, the areas of eruptive sun-spots outward from his equator.

While the nebulæ are more distant than many of the discrete stars revealed to us by the telescope, there is no reason to suppose that they are more distant than the star-clouds into which are merged the separate stars of the Milky Way, or the star-clusters seen in other portions of the sky. We know, in fact, that this is not so, for our telescopes show brilliant stars in very many cases which are components of the nebulæ themselves; and the fact that the nebulæ can be seen as having visible form, and not as mere points of light, is itself conclusive as to their relative distances. Hence we need not be surprised to learn that these forming spirals will result each in the production of a single solar system, and not a galaxy of suns, as was once supposed. Were such the case it would be impossible for us to observe the structure of the nebulæ at all, as their distances would be far too vast. Of the forms of the gaseous nebulæ Guillemin asks, "Is the spiral the original form of those gaseous matters, the condensation of which may give, or has given, birth to each individual of this gigantic association?" The same author says of these apparently regularly formed nebulæ, "It is impossible not to recognize in them so many systems." Many of the spiral nebulæ were formerly supposed to be globular aggregations of nebulous matter only, and their spiral character came as a great surprise with the use of more powerful telescopes; and many—nay, most—of these apparently globular nebulæ have totally changed their appearance when viewed with instruments of higher power, while the spirals have become more and more pronounced in character with every increase of telescopic vision. Of one of such apparently globular nebulæ Guillemin says, "The center is like a large globular nebula with a very marked condensation, whence radiate branches arranged in the form of spirals. In several points of these branches other centers of condensation are noticed. Sir John Herschel had classed this among the nebulæ of rounded, globular form, *doubtless because the central nebulosity was the only one revealed by his telescope."* The formation of the sub-centers in this nebula (which is between the Great Bear and Boötes) should be particularly noted in connection with the coalescence of planets as above described. In a note to Guillemin's work, Professor Lockyer says, "The proper motion of nebulæ has not yet been inquired into, because everybody, looking upon them as irresolvable star-clusters, thought them infinitely remote. Now, however, that we know they are *not* clusters of stars, properly so called, it is possible that they may be much nearer to us than we imagine."

In connection with the double-sun spiral nebula shown in the preceding illustration, Guillemin says, "We have noticed nebulæ accompanied by systems of double or multiple stars, placed in a manner so symmetrical in the midst of the nebulosity that it is impossible to doubt the existence of a real connection between the stars and the nebulæ." And Flammarion says of these apparently globular nebulæ, when under the observation of more powerful telescopes, "In the place where pale and whitish clouds gave out a calm and uniform light, the giant eye of the telescope has discerned *alternately dark and luminous regions,"—*that is to say, they reveal the operation of the opposite forces of attraction and repulsion, and are spiral. While gaseous nebulæ may be of any conceivable form, the direction and operation of the forces which will determine their character as solar systems must be similar, just as with the forms of organic life, and the only nebulæ which reveal a distinct systematic development in harmony with a working solar system are the spiral. There is no difficulty whatever in tracing such a nebula through all its formative stages, as we have done, and we can, in fact, see painted on the background of the sky every step of the shifting tableau through which such forms must pass.

By the nebular hypothesis the whole course of development, of necessity, is rigidly forward to its culmination; but by employing the analogies presented to us in other operations of nature, we can readily account for variations, haltings, ineffectual efforts, uncompleted processes, and even reversals and redistributions into other secondary sources of energy. They equally comprise the agencies for the production of a single solar system or of a myriad, just as we see the vortical water-spouts or sand-storms either single, double, or multiple; they are flexible, as are all the processes of nature, and require no violent assumption of a prior physical basis known to us " ne'er before on sea or shore." They also account for the deviation from the normal of the orbits of Neptune and Mercury, for the formation of the asteroids and Saturn's rings, for the different eccentricities and inclinations of the orbits, for the forward axial rotation of the planets and their satellites, and even for their perturbations and abnormalities; they furnish a basis for Bode's empirical law, for the distribution of the planets in size, for the origin of comets and meteor streams, for Kepler's laws, for the equal and permanent relation of eccentricities and inclinations, and for the fixed axial position of the moon with reference to the earth; they account for the free oxygen in the planetary and free hydrogen in the solar atmosphere, they employ the variation of volume of the sun as a regulator instead of an independent generator of light and heat, and they are in entire conformity with the established principles which govern the electrical generation of active forces, their transmission to the sun, their transformation into light and heat, and their return to the regions of space, where they continue to act with potential energy to all eternity, as they must do if space itself is eternal; and we surely know that, if anything whatever is eternal, space must be so. This great ocean—the home, the domain, the workshop of creative energy—is the last retreat of the human intellect; here it may find rest, and here alone. While solar systems may afford in their circling planets a possible dominion for finite life, and in their suns their daily bread; in the infinite and all-embracing realms of space, filled with the potentialities of all created forms, thrilled with the impulses of all creative force, is to be found the unfailing source of all, the dominion of the eternal architect, before whom nature bends the obedient knee, waits to hear his mighty voice, or swiftly runs to do his royal bidding.

CHAPTER XIV.

THE MOSAIC COSMOGONY.

"One generation passeth away, and another generation cometh: but the earth abideth for ever."—*Bible.*

THUS, as we have seen, through countless future ages will the sun, with his incandescent envelope of hydrogen, and the planets, with their life-sustaining atmospheres of oxygen, fulfil their appointed times and courses. But if we could conceive that all atmospheres, solar and planetary, were suddenly blotted out and forever annihilated, so that these great orbs thenceforth rolled along as they do now, but only as black globes in an ocean of space of Stygian darkness, new atmospheres would at once begin to be formed, and these would soon again surround the sun and planets, precisely like those which now exist.

Sweeping along in darkness, the force of gravity would gather around each of these bodies vast accumulations of aqueous vapor and other gases condensed from the attenuated matter of surrounding space. The planets, by their axial rotations, would again generate from these regions, newly occupied as the system drifted along through space, electrical energy of enormous quantity and potential. Earth would again hear the mighty mandate, "Let there be light," and from her poles to her equator the skies would blaze with brush-light auroras. Suddenly, with a mighty leap, the pent-up currents would flash across to their opposite electric pole, the auroras would gradually die away, and instantly the molecules of hydrogen would begin to sift out at the solar and those of oxygen at the planetary terminals. The electrical currents driving their furious pathway through the rapidly gathering hydrogen envelope, the sun would first begin to faintly flicker with hazy, nebulous light; the light would gather intensity, and soon flash and glow with energy; the solar nucleus within would become intensely heated and liquefied or partially volatilized, and again the solar streams of incandescent heat and light would radiate forth on every side; the commingled gases, oxygen and nitrogen, would once more surround each planetary globe, and we should have a new solar envelope just as we now see it, and new planetary atmospheres like our own; and then, and not till then, would the opposing generative forces permanently counterbalance each other and electrolytic decomposition become practically stationary, except to compensate for the slight variations constantly liable to occur in the complicated running of the mechanism. So the mutilated crustacean re-grows his lost claws, and so our own gaping wounds are healed by the great *vis medicatrix naturæ*. The most stable of all things is mutually balanced instability; perhaps there is no other form of stability.

The "Nebular Hypothesis" of Laplace concerns itself only with the aggregate matter of which our solar system is composed, and the force of gravity, including cohesion, ignoring the action of the equally powerful force of repulsion. But there is another nebular hypothesis much older than that of Laplace and far more scientific, for it utilizes both the force of gravity and cohesion and the radiant force of repulsion in the generation of our solar system. We refer to what is known as the Mosaic cosmogony. Whatever the origin of this magnificent narrative may have been, whether written down by Moses originally, or by him derived from the sacred learning of Egypt, with which he was fully acquainted, or by the Egyptian scribes drawn from Ethiopia, and still further back from the sacred traditions of India, it bears internal evidence, when properly rendered from the Hebrew record, of a knowledge of these stupendous phenomena (which no human eye could ever have beheld) which is most remarkable. The commonly accepted versions do not clearly bring out the full meaning of the original,—indeed, it would have been impossible for the earlier translators to have done so,—but when critically and etymologically rendered, very surprising coincidences with the succession of events as they must actually have occurred, and the principles involved in the successive stages of creation, will be found in nearly every part of the record.

This record is embodied in the first chapter and first three verses of the second chapter of Genesis. The Hebrew was long believed to be an original, if not an inspired, language, but it is now well known to have been a derivative or root language, made up much like the English, and, like it, having the meanings of its words primarily determined by those of the root-stems from which they have been formed. The roots of these Hebrew words are to be found among the languages of many older peoples, and nearly all of them have now been traced to their immediate origin. Another source of error is in the so-called Masoretic pointing, which was not introduced for a thousand years after the time of Moses, and which has often changed the signification of the older words, and even the form of the words themselves; but by critical researches the roots and their combinations have been isolated, so that we are now able to possess a much more accurate knowledge of the Mosaic record than was possible in former times, for, of course, no original copies have come down to us. It is not a reconstruction of the record which has been made, but a careful editing by means of the derivation and true signification of the words used, and by careful comparison among the most ancient versions accessible to modern research. The English version, while imperfect in its rendering of this ancient narrative, is not to be considered by any means a false translation, but it largely errs in failing to give the full radical meaning of the words employed in the original.

As an illustration of this indefiniteness of rendering in the ordinary English version let us consider the opening sentences of the narrative : " In the beginning God created the heaven and the earth. And the earth was without form, and void; and darkness was upon the face of the deep."

In the "beginning" of what? Does it mean the beginning of our own solar system? or of all systems? or of all space? or of Jehovah (for He has not yet been mentioned or described)? or of the Aleim themselves,—that is, did the work begin as soon as the forces began? and did the latter originate spontaneously, or otherwise? What "God" is meant? Is it Jehovah, or Aleim, or some other God not yet mentioned or described? If we will take every name in the Bible which is translated God (and it may be any of these according to the English rendering), we will have legion. We shall even find that the same word which is translated "God" was applied by Jehovah on one occasion to Moses. "Created"? What is meant by this word? Was the creating a creation out of nothing? out of something pre-existing? or something coexisting elsewhere? Was the creation a direct or an indirect one? by the use of the forces of nature, or by overriding the forces of nature? Was it a physical creation by an inconceivable action of mere thought, or will? and if so, was this thought, or will, God himself, or one of his attributes or powers only? "The heaven"? What heaven? Was it that to which the virtuous are supposed to go after death? or was it some more physical heaven? Was *the* heaven the atmospheric heaven, the interplanetary heaven, the heaven of interstellar space, or that more extended heaven which lies beyond our knowledge? Was *the* heaven one of those which He created, or did He create all the different heavens of all the solar systems and nebulæ at the same time? " Without form"? Was the earth without any form at all? or merely without its present form? or without some particular form not mentioned? If the earth was a physical structure it must have had *some* form; what was it? "And void"? Was the earth void like a soap-bubble? or void like a ray of light? or a vacuum? If it was empty, what was it that was empty? How could the heaven and earth be void after they had been brought into existence? " Darkness was upon the face of the deep"? What deep? Was it the sea not yet created? or the earth, which is anything but a " deep"? was it the atmosphere? or all space? If the latter, did all other systems of space wait for their light on ours? or did we wait on theirs? are there no new systems now forming, and none to be formed hereafter? If all space is meant, where was its outside, or its face? and what occupied the intervening regions? was it a physical face or the face of a vacuum? Were these statements to be accepted by faith or reason? If the former, was it a faith which could only have come from the experience of after-ages? or was it based on the *ipse dixit* of Moses? What was the basis of faith when the record was first written? was it from generally accepted tradition or by revelation? Is the record anonymous or does it reveal the name of its author? If to be endorsed by knowledge and reason, why should not the narrative be strictly and accurately translated, even at the expense of conciseness and elegance of diction, in order that the exact force of every word shall be fully felt and recognized? If the record is from divine revelation, it is still more essential to know precisely what was revealed; otherwise we are no better than idolaters; we are worse, in fact, for we have changed and falsified the landmarks of religion, and bear false witness against God Himself. We must not interpret Genesis by records made long subsequently; it must speak for itself or not at all.

When construed in accordance with the exact definition of the words themselves quite a new and strange light is thrown upon the history of the events thus recorded. The great importance of a strict construction of the translation and fidelity to the original is emphasized by the fact that the same word was never used in this record to express a different sense in different parts, nor were two different words ever used in different places to express the same meaning. It is, therefore, necessary to give every word of the original its exact fulness and force. The basis of the following critical translation is to be found in "Mankind: their Origin and Destiny" (Longmans & Co., London, 1872), but a careful comparison has been made with other accepted authorities, and the root-meanings of the separate words have been carefully traced out, so that many necessary changes will be found to have been made in order to bring out the precise sense of the original. There is no actual literal, critical, etymological, and scientific rendering embraced in a single translation known to us, and which is complete in itself; but that which follows will be found, it is believed, to give every word its particular etymological shade of meaning, and to employ the same word in the same place, for the same purpose, and with the same signification as it was understood to have, in its original form, when first recorded. The specific root-meanings of the most important words used are further explained in detail in a separate section below.

The use of ALEIM, "the powerful Forces," in the plural, followed by the verb in the singular, is a Hebraism, and indicates the collective character of the forces as specially energized, sent forth, and directed by Jeove (Jeova or Jehovah is the Chaldaic form of the word, the original Hebrew being Jeove), who does not appear by name in this narrative, though, as we shall see, specially delegated power from some higher source is that characteristic which is most emphasized throughout the record. These forces are personified, as is usual in ancient records (and, indeed, in modern thought), but they are in reality the "powers of God." The author of the work above referred to says, "The idea of Moses was that there was a Supreme God . . . and that He only acts by means of his agents called ALEIM, the Gods, in the plural and indefinite number, or embassadors, or voices." The ancient belief in the unity of all forces in one creative individuality is also most clearly shown in some of the oldest Vedaic hymns of India (see Max Müller, "The Veda"). "Self (Atman) is the Lord of all things, Self is the King of all things. As all the spokes of a wheel are contained in the nave and the circumference, all things are contained in this Self; all selves are contained in this Self. Brahman (Force) itself is but Self."

Of the religion of the ancient Egyptians (see "Evolution and Christianity," by J. F. York) it is said, "The chief theological characteristic of this first of all known civilized religions is the doctrine of the Divine Unity. As M. de Rougé says, 'One idea predominates, that of a single and primeval God; everywhere and always it is one substance, self-existent, and an unapproachable God.'" The Egyptian cosmogony, as the fragments have come down to us (see Professor Arnold Guyot, "Creation"), is as follows:

1. The *original gaseous form, and the darkness of matter*.
2. The successive transformations.
3. Light, as the first step in this development.
4. The separation of the waters below from the waters above the expanse.
5. Periods of development of indefinite length.
6. The sun, moon, and earth organized last.

The word MLACTOU, which occurs several times repeated in the summing up of this narrative, explains the character of ALEIM most fully, as specially energized and directed agencies or forces. This word never has any other meaning. Even when applied to a king it was not a king as a monarch, but as the specially directed agent of God. I. Samuel xxviii. 17, "The Lord hath sent the kingdom out of thine hand; . . . because thou obeydst not the voice of the Lord." When, in Exodus xiii. 21 it is said that "Jeove went before them by day in a pillar of a cloud," this is explained, in chapter xiv. verse 19, to mean that this pillar of cloud by day and of fire by night was MLAC, a messenger, or agent. It is translated "angel" in the English version, but it was not a personal angel; it was a specially energized and directed force. In the earliest times it was not the God of fire, or of force, or of justice which men feared, but fire, or force, or justice; the anthropomorphic conception came later with the generalization of all fire, all force, or all justice. We say now that a malefactor fears the law; what he really fears, however, is punishment. In this record we are dealing with the primordial forces of God,—gravity, electricity, attraction, repulsion, cohesion, vital force, etc., etc., but acting with special energy for a predetermined result. Of these forces Dr. McCosh says, in his work on Christianity and Positivism, "One God, with his infinitely varied perfections,— his power, his knowledge, his wisdom, his love, his mercy; we should see that one Power blowing in the breeze, smiling in the sunshine, sparkling in the stars, quickening us as we bound along in the felt enjoyment of health, efflorescing in every form and hue of beauty, and showering down daily gifts upon us. The profoundest minds in our day, and in every day, have been fond of regarding *this force*, not as something independent of God, but as the *very power of God acting in all action;* so that in him we live, and move, and have our being." In more rugged and virile form this was precisely the old Mosaic philosophy, the philosophy of the arcana of the Egyptian temples, and of the Vedaic age of the Aryans of India. Where was the radiant center of this unfailing search-light which has poured its broad belt of dazzling brightness down to our day from those old, prehistoric ages?

So De Jouvencel, in his "Genesis according to Science," says, "We should not place the works of nature on one side and nature on the other. Nature is a work and not a person."

The word which in the English version is translated "rested," in the concluding verses of the narrative, does not mean *rested from fatigue*, but rested as a pendulum rests when it ceases to vibrate. Had the word been rendered "came to a state of rest," it would have been far more accurate and true to the sense of the original. What is meant is that these pent-up forces had operated, under the guidance of Jeove, to rupture a state of unstable equilibrium in the attenuated matter of space, just as similar forces are now said to gather energy to produce a volcanic eruption of the earth's crust, preceded by earthquakes and other vast disturbances radiating from the center of rupture of these tensions between the molecules of matter, accompanied by explosive expansion and all the phenomena of disorganization and repulsion, and succeeded by condensation, development, harmony, and final quiescence of these specially energized and self-opposing forces in a newly formed state of molecular equilibrium. To quote from Professor Guyot, "God rests as the creator of the visible universe. *The forces of nature are now in that admirable equilibrium* which we now behold, and which is necessary to our existence." In "The Unity of Nature" the Duke of Argyle says, "We strain our imaginations to conceive the processes of Creation, whilst in reality they are around us daily."

The words which conclude the third verse of chapter ii. are also imperfectly rendered in our English version, and this defect has led to a popular misconception almost universal. They are construed to mean "created—and made," as though marking a broad class distinction between the difent processes before described. From this the inference has been drawn that while, for the more subordinate features, the word rendered "made" indicated that these were stages in the process of creation merely involving the use of coexisting materials, in the grander features of the work it was supposed that there had been a creation *ab initio*,—that is, *out of nothing*. Whole libraries have been written on this theme; but the words used bear no such meaning; on the contrary, they signify the exact opposite. There is, however, a broad distinction between the interpretation of the two words; but it is that the word which is to be rendered "fashioned like the work of a sculptor" is narrower and not broader in significance than the simple word "made;"

so that the former is included in, but is not generically distinct from, the latter. The word BRA means that these portions of creation were fashioned with the care and artistic skill of a sculptor, as contradistinguished from turning out the productions in mass; this distinction does not relate to the origin, but to the workmanship. However interstellar or primordial space was formed, or when, if it ever was formed, there is nothing in this record which excludes a pre-existent space substantially like that which now is. What we see in the sky, among the nebulæ, are later developments of like solar systems, in like manner, from the midst of the substance of the same illimitable and eternal space.

But biology has an interest in this account of creation equally as great as has cosmology. The word BRA is first applied to the formation of the individualized substance of the heavens and the earth. They were fashioned or carved out like a sculpture from something on which the forces could operate. There was, of course, *creation* involved, but it was a mental, not a physical process. When a sculptor has completed his clay figure he has brought forth a *great creation*, perhaps, and the "creation" is still his own, though the figure be cast in bronze by hired workmen in the foundry, who execute the sculptor's will at two dollars a day, it may be, each. Beyond this mental element there is no more *creation*, in its widest sense, than when a boy "creates" a new point on his pencil by guiding his hand and knife to sharpen it.

When the "diffused light" came, it is not said that it was "fashioned like the work of a sculptor," or that it was even "made;" but that it "came into existence." "Let there be light, and there was light," as the English version has it. But when the radiant energy of the sun came to be formed, on the fourth day, it did not "come into existence," nor was it "fashioned like the work of a sculptor;" it was "made." The reason is that it was not a development from the preceding "diffused light," but a new kind of light, made mechanically by the electrolysis of aqueous vapor around the sun's body, forming a hydrogen envelope, and by driving the furious torrents of electricity from the planets through this atmosphere, while the auroral, "diffused light" of the earth was gradually dying away during the process. Hence there was no room for the word BRA, or for the word IEI (came into existence) here; the word to be used was OSH. And when life was first introduced,—vegetable life, the primal life,—the word used is not BRA; this life was not "fashioned" or developed from other life. But when animal life was afterwards introduced, the word used is BRA; it was a refashioning. What was this life fashioned out of? It was not "made;" it did not "begin to exist;" it was developed. In this manner the earth was finally filled with animal life. Then came the introduction of the human race. Here we again have the word BRA, thrice repeated; but when this introduction of mankind was first projected, and before it was executed, it was in these words, " We will *make* [the root OSH] mankind ;" or, in the English version, " Let us *make* man." There seems here to have been a gradual ascent of living organisms by development, almost precisely in accordance with the most recent teachings of science. Two essentially different *kinds* of light were successively produced, independently of each other; the earlier kind " came into being," and the later " was made." The substance or entity of the heavens and of the earth, generically, " was fashioned." Three successive introductions of organic life not essentially different from each other occurred; the first is described thus : " Let the earth bring forth ; . . . and the earth brought forth," in the English version; or " There shall be made to grow ; . . . and there was caused to arise suddenly out of the ground . . . vegetation," as more accurately rendered. The second form of organic life, in order of time, the animal, was " fashioned." The third form, mankind, was also " fashioned," and this was done long subsequently to the introduction of the second.

If the word BRA had any signification of *original creation* it would have been applied to the first creation of life, for it was far more wonderful and original that there should be vegetable life which grew and developed, which brought forth flowers and then fruit, which formed germinative seeds, and from these successively and continuously reproduced its multifarious species, than that *animal* life should have been introduced long afterwards to repeat these same things which vegetation had been, in all its forms, from the lowest to the highest, already doing for untold ages,—from the third period of the earth's long history to the fifth; and more especially still when we consider that vegetable life and animal life, in their lowest forms, have no positive line of division between them.

And if OSH, which is applied to the genesis of solar light, be capable of the signification of *original creation*, then this word should have been applied to the generation of the "diffused light" of the second day, for the genesis of light is far more wonderful and original than the subsequent production of sunlight, after the forming earth had existed for two whole formative periods, from the second to the fourth, under the constant illumination of this universally diffused auroral light. If, on the other hand, the words applied to the first generation of light and the first generation of life be held to mark an *original creation*, then these words are never applied in this whole narrative to the genesis of the *entity* of the heavens, or the earth, or the sun and moon, or to animal life, or the life of man.

The radiant light and heat of the sun were not made until the fourth day, while the introduction of vegetable life dates from the long antecedent third day of creation. Prior to the development of the sun's thermal light there could have been, as we have already shown, no free oxygen in the terrestrial atmosphere; and it is a remarkable circumstance that vegetation, which is the only form of organic life which could have existed and propagated its species in an atmosphere composed of carbonic, nitrogenous, and aqueous vapors, devoid of oxygen, is that particular form of life which has been selected for this purpose, and its advent placed prior to the making of the sun. It would have been far more reasonable (previous to our present knowledge of these things) to have placed the formation of the sun in advance of the introduction of life; it is surprising that this was not done, unless we give to these "ancients" a knowledge of the principles of natural science far beyond anything hitherto attributed to them.

In the same connection there is described a stage preparatory to and leading up to the simultaneous development of the sun's light and heat, and the sifting out of hydrogen around the solar core, and of oxygen in the terrestrial atmosphere, which is equally remarkable. The "separation of the waters" described in verses 6 and 7 has never been fully rendered into English, or even understood in the original, as the words seemed meaningless in their literal sense until correctly interpreted by the facts set forth in the present work.

We must first note that the separation of the waters of space to two opposite foci, with an intervening space of attenuated matter, and their condensation there into two entirely different bodies, was the work of the second day, while the formation of the terrestrial rain-clouds and seas, as connected together, was a work of the third day, and was not accomplished until then, which was long afterwards. These entirely different operations— different in time, place, character, and circumstance —have always been confounded with each other; but one is in reality systemic and the other merely local.

In verse 6 there was decreed an expanse or *thinning* (an attenuated region) in the *center* of the waters, and a separation was made by the formation of two "spots" (verse 7), one under the expanse and the other above the expanse; the expanse was space, interplanetary space. Professor Arnold Guyot, in his book on Creation, says, "It is to be regretted that the English version has translated the Hebrew word *expanse* by the word *firmament*. . . . The difficulties they [the commentators] have created for themselves arose . . . from depriving it of its cosmogonic character and belittling it by reducing the great phenomena there described to a simple modification of the terrestrial atmosphere. . . . They forget that this thin covering of clouds is but a temporary and ever-changing one, and that the clouds are *in* that heaven rather than above it. . . . They forget that this is not the true heavens in which are spread the sun and moon and stars. . . . This grand day, so dwarfed and misunderstood, is the one in which are described the generations of the heavens, announced by Moses, which otherwise find no place in the narrative of the creative week."

The two foci of waters were the solar and terrestrial; around these bodies were gathered by the attraction of gravity, and there condensed, the aqueous vapors from the attenuated intervening matter of space; the earth by its rotation generated the enormous electrical currents which still continue; when these made their mighty leap across to the sun, the diffused auroral light around the earth gradually disappeared, hydrogen and oxygen began to be evolved at the opposite poles—the sun and the earth—from the condensed envelopes of aqueous vapor which surrounded them, the sun's hydrogen atmosphere was pierced, as in the pail-of-water experiment described in an earlier chapter of the present work, by the planetary electric currents, the sun became incandescent, and *pari passu* the earth became fitted, by the development of oxygen, for the abode of animal life. As taking part in this great mechanical transformation, the sun was said to have been "made;" it did not "come into being."

Just prior to the introduction of vegetable life—during the same creative epoch, in fact, and for the support of which life it was necessary—the waters under the expanse were condensed into rain-clouds and seas, and there is a curious reference (verse 9) to the appearance of the earth's dryness "as produced by the action of an internal fire;" the gradual cooling of the earth by the radiation of its internal heat of condensation into space would account for this appearance, and, in connection with the diffused auroral light throughout the whole sky, would doubtless have sufficed for the support of vegetable life.

In verse 16 the fixed stars (the suns of other systems) are referred to, but in a parenthetical statement—almost deprecatory, in fact—that "the dim and almost extinct lights" the same forces created also, but when they were created is not stated in the record. The occasion for this incidental remark is to be found in the preceding statement that the two new luminaries, the sun and moon, were the two "superior bodies in size of the starry lights." Having mentioned the stars in this comparison, the author feels called upon to add that the latter also had been similarly created,—that is, that they were not original existences, and of course they are not, but they were not created at that epoch, and are not said to have been.

In chapter ii. verse 4, which opens the second narrative (quite a different history, by the way), Jeove appears Himself, joined with the Aleim, and henceforth this personal connection is maintained; the English version translates this composite word "The Lord God," which means the Master God; the correct reading is, however, the "God of gods," or what we call the "God of the forces of nature," or the "God omnipotent."

In the whole Mosaic cosmogony there is nothing which can even suggest a gradually closing nebulous mass; the element of rotation is absent (and it would not have been understood by the people even if presented); but, with this exception, the processes of development are substantially in accord with what must really have taken place, and in the order described. But it is, as before stated, absolutely essential to understand the root-meanings of all the more important words used in the original. A superficial translation is not only meaningless, but misleading; whereas, when accurately understood, the record is one of the most remarkable ever presented to human intelligence. The words used were selected deliberately for their specific shades of meaning, and, unless these are properly rendered, to the uninformed the narrative will present a simple succession of startling phenomena, while to the educated student each of these changes carries within its verbal index its origin, its mode, and the knowledge of the forces at work. To the one it is a dramatic spectacle performed on the stage in front; to the other it is the same work as seen behind the curtain, with all the intermoving mechanism (the author's manuscript the sole guide), the interplay of complicated forces, the triumphant successes, the rapt attention, and even the sudden applause extorted at each wondrous climax from the skilled actors themselves, who are at the same time unceasingly engaged in working out the mighty drama of creation. One might readily believe that the original author of this record was thoroughly acquainted with the processes involved in the development of a solar system like our own from the diffused primordial matter of space, substantially as we have endeavored, in the present work, to deduce them from the most recent investigations and discoveries of science.

Of the watery vapors condensed above the expanse of space many of the ancient writers had a far more correct knowledge than had those who translated these chapters from the original into the various modern languages. In the Psalms we read, "Praise him, . . . ye waters that be above the heavens;" in the Song of the Three Holy Children, "O all ye waters that be above the heavens." Theophilus speaks of the "visible sky as having *drawn to itself* a portion of the waters of chaos at the time of the creation. Saint Augustine says that the firmament has been formed "*between* the upper and the lower waters," and quotes Genesis i. 6 and 7, as his authority.

Thousands of years ago, as far back as the days of the Pythagoreans, and even long before, mankind was acquainted with the mariner's compass, telescopic tubes, and glass lenses; they knew that the moon receives her light by reflection from the sun, of the presence of mountains and valleys on the lunar surface, that her day and night are each a fortnight in length, that there were other planets known to the Egyptians besides the seven known to the Greeks (the Brahmans reckoned fifteen of them), that the sun is the center of our planetary system, that the earth and the other planets revolve around it, that the earth is round and rotates on its own axis daily, that weight is a principal element in the maintenance of these rotations, that the fixed stars are suns, and that the Milky Way appears white from the number of stars which it contains. Kircher quotes from an ancient Syrian author the philosophy of the sidereal system, dividing it into many layers or spheres attached to orbits, each presided over by a spirit. In the eighth sphere are placed the fixed stars, "still higher two other layers of stars not less luminous, and of different sizes, the nebulæ and the small stars of the Milky Way, and the whole is surrounded by the celestial waters, which spread over the whole firmament, and which compose the great sea of light and the boundless ocean." The sources of all this wondrous knowledge can be traced back through Chaldea, Arabia, Egypt, Ethiopia, and, through the colony of Meroë, to India.

ROOT-MEANINGS OF THE PRINCIPAL WORDS USED IN THE MOSAIC NARRATIVE OF CREATION.

ALEIM ("corruptly called Elohim by the modern Jews, but always Aleim in the synagogue copies") means the Strong Forces (or, by subsequent impersonation, subaltern gods), operating to carry out the purposes and execute the plans of Jeove. AL, the root, signifies *Strong, strength, a ram;* AL-E means *Strong* in a personal sense; ALEIM (plural) means the Forces, the Strong-ones, the Powers, and in Egyptian mythology, the subordinate, or executive, gods, the demi-urgi. Exodus vii. 1, "And the Lord [Jeove] said unto Moses, See I have made thee a god [Aleim] to Pharaoh; thou shalt speak all that I command thee."

BRA, *carved, cut, fashioned like the work of a sculptor, gave a new shape to, formed from unformed material.* From BR, *a knife;* BR-I, *to carve, to cut.*

BRASHIT, *in the commencement* or beginning *of individualized existence* (with the initial preposition B-). B signifies *in;* IT (which is related to AT) signifies *individualized existence;* RASH, a *principle* or *beginning,* or a *commencement.*

AT, connected with the Chaldaic, signifies *substance, essence,* or *individuality,* "the thing itself" (Latin, *ens*); it is correctly translated "individualized substance."

ESHMIM, the combination of the preposition E with the substantive SHMIM, the word signifying *of the visible heavens*, or the planisphere.

ARTZ, the earth in a state of aridity, or as a generalized expression for the earth; AR signifies the *earth*, and the termination TZ intensifies the signification of *drought, whiteness, aridity;* in contrast with this is ADME, *red earth*, or *productive earth* or *soil*.

U– is a conjunction, signifying *and* or *then*, in the sense of succession of time, something like our phrase "and then."

TEOU does not mean "without form," nor does UBEOU mean "and void," as rendered in our English version, at least not in the ordinary sense of these words. "TEOU refers to extinct life, or to existence *shut up as in a tomb and in darkness*, while U-BEOU refers to *life which is about reappearing*, but still hidden in the egg or the ovary, and waiting for the word which shall cause the dawn of creation to shine upon it." These words are more properly rendered "tomb-like darkness and undeveloped."

ESHC means *darkness;* not merely an intense darkness, but what may be denominated a "thick darkness;" it is an *enshrouding darkness* which *compresses and hinders*. It is precisely such a darkness as would be produced by the interstratified cloud-layers between the convolutions of a forming spiral nebula, or the cloud-strata surrounding the earth before electrolytic decomposition of the aqueous vapors had ensued. With the advent of the sun, in the narrative, this darkness and the term which expresses it disappear.

TEOU-M is the word above explained, with the termination –M, expressing the idea of *arrested, doubtful, indefinite*, as applied to all existence; the word "undifferentiated nature" properly interprets its vagueness and general character of an abyss of being, in the etymological sense of "nature" as the totality of things at that time born or produced.

ROVE means *breath*, in the sense of an expanding, liberating, or developing spirit; its literal meaning is "the breath, the spirit which dilates and frees."

MREPHT, *brooded with incubating love;* REPH is composed

of RE, "to be full of good-will, to be agreeable," and EPH, "to cover, to protect, to incubate, to brood."

MIM, *the seeds of all beings, the waters.* It is said, "the choice of this letter M, to signify water [the alphabetical Egyptian letter M is represented by the two undulatory lines which in the hieroglyphics represent water], is connected with the Egyptian ideas of the cause of the generation of living beings." Numbers xxiv. 7, "He shall pour the waters out of his buckets, and the seed [ZRO] in the waters [B-MIM]." The latter word is plural in form, but both singular and plural in sense.

AOUR, *diffused light;* a light resembling the dawn, but quite distinct from the light of the sun. The latter was not established until the fourth day, and its advent is characterized by a new word, LEAIR, "to cause light to *move* above the earth."

JOUM is *day,* generically, and LILE *night.*

RQIÔ, *the expanse;* ATRQIÔ, *the individualized substance of the expanse.* Space, in the opinion of the Egyptians, "not being a vacuum, but a material substance, Moses could say, and was even compelled to say, 'the substance of space, that which constitutes it.'"

OSH, *made.* This word first occurs in verse 7, and is there applied to the *making* a separation between the waters or aqueous vapors condensed around the earth and those condensed around some similar spot "above, as regards the individuality of the expanse,"—to wit, the solar core or nucleus, —to which, attracted by gravity from the attenuated vapors of the space between, is due the subsequent establishment of the solar light and heat, as in an electrical arc light, and the presence of oxygen in the terrestrial atmosphere. These processes, involving the constitution of our atmosphere and of the sun's photosphere and chromosphere, were not completed until two subsequent cosmical periods had elapsed, from the third to the fifth. The word OSH, in its different combinations and inflections, is also used in verse 11, where it signifies "making," as applied to fruit; "yielding" fruit, in verse 12; "they made," as applied to the sun and moon, in verse 16; "made," as applied to the entity of

THE MOSAIC COSMOGONY. 333

quadrupeds and higher animals generally, in verse 25; "we will make," as applied to man, verse 26; "had made," as applied to "every entity of creation," verse 31; "had made," as applied to the specially directed work as MLACTOU, chapter ii. verse 2; and finally, in the general summing up in verse 3 of the second chapter, as an element in a compound substantive phrase "according to the making-act," or "in accordance with the making of creation."

"OSHOUT," it is said, "signifies a manual operation, carried on according to a previously conceived idea, or model."

We find a similar use of the substantive infinitive with a preceding preposition in verse 21, chapter iii. "CTNOUT is derived from TNE, a consoling word. TNOUT, the infinitive of the conjugation Piel, adds to the word the act of causing to be done, and of doing with care."

A similar construction, LRAOUT, is employed in chapter ii. verse 19, translated in the English version, " and brought them unto Adam *to see* what . . ."; more literally, "as regards the act of seeing," or according to a vision, or show. That is, they were brought and presented to his sight.

The object in writing these two words, BRA and L-OSH-OUT, together at the very end of the narrative was to conclusively establish the fact, beyond all possible doubt, that the whole work of creation was an orderly and harmonious progression.

MLACTOU, which word is used twice in verse 2 and once in verse 3 of the second chapter, and not previously, is also introduced for specific emphasis. It means that the whole preceding work of creation was, in its nature, "the work of Mlac," a messenger, or a specially energized and directed agency, sent to fulfil the appointed work of Jeove. Its purpose was to forever prevent the belief that the work of creation was due to mere natural forces, on the one hand, operating by chance; and, on the other, that these forces were independent gods carrying out their own purposes, and of their own will. It was set up as a double barrier against rationalism on the one side and polytheism on the other.

It may be incidentally added that the popular belief that

"Adam was created out of the dust of the earth" is not in accordance with the original record. In the second narrative, chapter ii. verse 7, the word OPHR is rendered "dust" in our English version, but it does not signify ordinary terrestrial dust at all; "its radical meaning is to volatilize a substance, to sublimate it." The true signification of the word used is analogous to a "material essence." The same word is used in Numbers xxiii. 10 as a synonym for "seed;" it is said that "the Septuagint version translates OPHR by *sperma*."

The formation, described in the third chapter, of the female human being out of one of the ribs of Adam, excised for that purpose (which is a matter of almost universal popular belief), is not, in reality, what is stated in the original. In verse 21 of chapter ii. the words are rendered in our version, "And he took one of his ribs." What is really said, however, is "And he brought out another one from his sides." So the similar expression in verse 22 in reality signifies, "caused to be made according to womankind the individualized substance of his side."

The word translated "*of his ribs*" is precisely the same as is subsequently used by the same writer (Exodus xxxvii. 27) to designate the location of the supporting rings upon an altar of incense, and is there rendered, "by the two corners of it, upon the two sides."

The defective translation is due to imperfect knowledge, at that time, of the processes of organic development. The true signification is that given in the "Institutes of Manu": "Having divided his own sub-sistence, the Mighty Power became half male and half female."

The words rendered "help meet" in verses 18 and 20 have a far higher meaning; "I will make him a help meet" should be translated, "I will cause to be made for him an overseeing help as a guide, an instructor, a revealer." And in verse 20 of chapter iii., "And Adam called his wife's name Eve," the latter word is not translated; the correct rendering is, "And Adam called the symbolic name of his wife the female serpent-wise revealer, she who explains, points out things, who instructs," for that is what the true root-meaning of Eve signifies. The concluding words of this verse, "because she was the mother of all living," are obviously mistranslated, for not only was she not a mother at all, but she did not even conceive, as stated in the next chapter, until she had left the garden finally. The true signification is, "because she was the mother of all [spiritual, see verse 22, as contradistinguished from animal and vegetable] life."

The female human being, the word translated woman, has the generic root-signification of "flame," while, prior to Eve, that of the Adamic man is the "red earth." As the male was formed from a material earthly essence, the female was created one remove further from the gross and material in the direction of the spiritual; and her powers were distinctively subjective, those of intuition, while those of the male were objective, those derived from instruction. Even in the final curse (so called) the man turns back to the earth to earn his subsistence, while the woman turns forward to the instruction of the future men and women, the children; for the words, "In sorrow shalt thou bring forth children," have left one word of the original untranslated, and by supplying this the sense is entirely changed, "and conceiving, and bringing forth, in sorrow shalt thou bring up, care for, and train children." In those countries childbirth was never attended with much pain or sorrow.

The obvious effect of the whole inspired or traditionary second narrative is to clearly differentiate the contrasted faculties of the two sexes, and the root-meanings of the words employed, whether Moses himself perceived it or not, are a testimonial of the highest possible character for woman, instead of being, as rendered in the ordinary versions, a mark of inferiority, or even of degradation. In the garden scene, when she partook of the fruit of the tree of knowledge, she did not do it hastily or from mere temptation; it is said that "she considered it attentively;" the same word being used as was employed in the first narrative to mark the intense interest and almost superhuman character of the consideration by the Aleim of the work, as its successive stages appeared, which they were delegated to perform, and which Jeove himself directed. The prize, to her, far outweighed the penalty, and the aspiring sibyl dared to lift the innermost veil in the adytum of the temple, and grasp the lofty truths which made her as one of the Aleim. So fell Prometheus.

And then, no sooner had the flame-crowned seer won her precious prize, than, woman-like, she turned and laid it before her husband, and he, the innocent one, "did eat."

The serpent was not a mere snake, be it understood; it was the Egyptian Typhon, the dark Spirit of doubt, the questioner, the tempter, the eternal IF, the why, whence, what, and whither?

It was her insatiable aspiration to reach the highest possible limits of human knowledge which gave strength to her daring, and not a childish fancy for an apple. All this, of course, is lost in the translation. It is as though the national standard of a mighty people had been disinterred from the remains of past ages, which had been borne aloft at the head of mighty armies for centuries, and for which thousands had gloriously died in battle in defence of a sacred cause, and which now, its past history untraced, has been catalogued as a brass bird of some sort mounted on a stick.

It is to be regretted that there is no plain, popular work by a thoroughly capable scholar, without theological or antitheological bias, which treats of the origin, form, root-derivation, usage, accurate signification, and construction of the comparatively few words employed in the ancient narratives which compose the first half-dozen chapters of Genesis, and, we may add, the book of Job; something like those inestimable works which deal with the ancient cosmogonic literature of Egypt, Babylonia, Persia, India, China, Phœnicia, and Central America. Nothing of this sort is to be found, at all events in a form accessible to the general reader, and such a work, in small compass, would be of the highest importance to popular instructors, to students, and to the public as well, for it would throw a flood of light on these extremely valuable but, hitherto, so illy-comprehended records.

THE MOSAIC COSMOGONY. 337

THE MOSAIC NARRATIVE OF CREATION.

1. ALEIM, the Forces, fashioned like the work of a sculptor, in the commencement of individualized existence, the individualized substance of the heavens and the individualized substance of the earth.

2. And the earth was in tomb-like darkness and undeveloped, and there was compressive hindering darkness on the surface of undifferentiated nature. And the dilating and liberating Spirit of the Forces hovered with incubating love on the surface of the seeds of all beings, the waters.

3. Then Aleim said, There shall be a diffused light; and a diffused light was.

4. And Aleim regarded with attention the individualized substance of the diffused light, because good. And Aleim caused a separation to be made between the diffused light and between the compressive hindering darkness.

5. Then Aleim exclaimed for the diffused light, DAY! and for the compressive hindering darkness exclaimed, NIGHT! And there was a transition from light to darkness, and then there was a renewal of light; FIRST DAY.

6. Then Aleim said, There shall be an expansion obtained by a thinning in the center of the waters, and there was that which caused a separation to be made by occupying a spot, the waters according to the waters.

7. And Aleim made the individualized substance of the expanse, and caused a separation to exist by the occupation of the spot, of the waters which are under as regards the expanse of space, and by the occupation of the spot, of the waters which are above as regards the expanse of space; and it was so.

8. Then Aleim exclaimed for the expanse of space, THE HEAVENS! and there was a transition from light to darkness, and then there was a renewal of light; SECOND DAY.

9. And Aleim said, The waters which are underneath the heavens will tend directly, in order to meet in it, towards a single spot fixed upon for their meeting; and of dryness produced by the action of an internal fire the appearance shall be made; and it was so.

10. Then Aleim exclaimed for the dryness, EARTH! and for the spot fixed upon for the meeting of the waters exclaimed, SEAS! Then Aleim looked attentively at it, because good.

11. And Aleim said, There shall be made to grow from the earth a dwarf vegetation which can be trodden under foot, a maturing plant causing to be sowed around it a seed, the strong and woody substance of fruit making fruit after his kind whose seed is in itself above the earth; and it was so.

12. And there was caused to arise suddenly and full of strength a dwarf vegetation, a maturing plant sowing around it seed after his kind; and the woody substance yielding fruit whose seed is in itself after his kind. Then Aleim considered it, because good.

13. And there was a transition from light to darkness, and then there was a renewal of light; THIRD DAY.

14. Then Aleim said, There shall be starry-lights in the expanse of space of the heavens to separate between the duration of the day and between the duration of the night; and they shall be for signs, and for seasons, and for the days which make the year, and for the repetitions of years.

15. And they shall be for luminous bodies in the expanse of space of the heavens to cause light to move above the earth; and it was so.

16. And Aleim made a double individualized substance, the superior in size and excellence of the starry-lights, the individualized substance which was the greater of the luminous bodies to represent the rule of the day, and the lesser luminous body to represent the rule of the night.

Of the dim and almost extinct lights [the stars] they made the individualized substance also.

17. And Aleim established these individualized substances in the expanse of space of the heavens to make light move above the earth.

18. And to be representatives of dominion during the day and during the night, and to separate between the continuance of diffused light and between the continuance of compressive

hindering darkness; then Aleim looked attentively at it, because good.

19. And there was a transition from light to darkness, and then there was a renewal of light; FOURTH DAY.

20. Then Aleim said, The waters shall bring forth a swarm of swarming creatures having living breath; and that which flies, the birds, shall be made to fly with strength and fleetness above the earth in the space extended of the heavens.

21. And Aleim fashioned like the work of a sculptor the individualized substance of those which are superior in size of the gigantic reptiles and every individualized substance having living breath, that moveth, which they had produced, swarming from the waters, according to their kind; and every individualized substance of flying thing with wings, after his kind. Then Aleim looked attentively at it, because good.

22. And Aleim blessed these individualities by saying, propagate your species and multiply yourselves, and fill the individualized substance of the waters in the seas; and as for the flying thing, it shall multiply itself on the earth.

23. And there was a transition from light to darkness, and then there was a renewal of light; FIFTH DAY.

24. Then Aleim said, From the earth shall be brought forth the living breath according to its kind, the quadruped, and the being which moveth about, and the terrestrial animal according to its kind; and it was so.

25. And Aleim made the individualized substance of the animal of the earth according to his kind, and the individualized substance of the quadruped according to his kind, and every individualized substance that moveth about of red earth according to his kind. Then Aleim regarded it, because good.

26. Then Aleim said, We will make mankind of a like order of intellect with ourselves, and they shall extend their dominion over the fish of the sea, and over the bird of the heavens, and over the quadruped, and over all of the earth, and over all the moving beings that move about over the earth.

27. And Aleim fashioned like the work of a sculptor the

individualized substance of mankind in the exactness of a shadow cast upon a wall; on this shadow Aleim carved the individuality; male and female they fashioned the individualized substance.

28. Then Aleim blessed the individualized substance. And Aleim said unto them, Be fruitful and multiply and replenish the individualized substance of the earth, and subdue it, and extend your dominion over the fish of the sea, and over the birds of the heavens, and over all life of the being which moveth about over the earth.

29. And Aleim said, Behold I have given for you every useful plant-substance yielding seed, yielding seed which there is over the surface of all the earth, and every individualized substance of tree which has in it fruit pertaining to a tree yielding seed, yielding seed for you, it shall be for food.

30. And for all animal life of the earth, and for everything that flies in the heavens, and for every being that moveth over the surface of the earth which has in it living breath, every individualized substance which is a green maturing plant shall be for food. And it was so.

31. Then Aleim looked at every individualized substance which they had made, and behold it was as good as possible. And there was a transition from light to darkness, and then there was a renewal of light; SIXTH DAY.

(Chapter ii.) 1. Then the finishing was made of the heavens, and of the earth, and of all the orderly arrangement.

2. And Aleim [the Forces] finished on the seventh day the divinely appointed and directed work which they had performed; and they came again to a state of rest on the seventh day from all the appointed work which they had done.

3. Then Aleim blessed the individualized substance of the seventh day and sanctified it, because in it they returned to their primitive condition from all the divinely appointed and directed work which the Forces had fashioned like the work of a sculptor, in accordance with the making of creation.

CHAPTER XV.

CONCLUSION. THE HARMONY OF NATURE'S LAWS AND OPERATIONS.

WE have passed before us the different orders of celestial phenomena; we have called down the denizens of the starry skies and placed them on the witness stand, and we have interrogated them in the light of the evidence which they have given before; we have compared their different statements, and have found that in their testimony they all finally agree. Instead of confusion, we find order; instead of complexity, simplicity; instead of discord, harmony; and through all we see the orderly progress of nature with uniform step, from stage to stage, higher and higher, until at last she stands triumphant, the handmaid of creative power, in the very center of the arch of the universe. We have taken the simplest operations which we find in progress around us, and have extended them to larger operations, constantly keeping in view their relevancy and the facts which form their sole support. Mere speculation has been excluded, and theory has found its every step based on an established fact. In this way we may hope to make place for further investigation in this field by abler minds, and that the conclusions of science may then become so well understood and so firmly established that to go back to the "dead-and-dying" theories of solar energies will be like going back to Ptolemy and Tycho for our astronomy.

We have considered the hypothesis which bases the energy of our sun upon his inherent heat, upon combustion, upon the accretion of meteoric streams, and upon his slow and gradual condensation of volume; and have found that all these hypotheses, singly or combined, fail to account for his energy through the vistas of the past, during which we know he must have shone as he now shines, and fail to account for more than a slow but inevitable decline, in the relatively near future, into eternal darkness and death. We have found that all these theories are alike, in that they recognize the sun itself as the only source of his energy, that his enormous emission of light and heat is almost entirely wasted in empty space, and that this will go on with the same frightful waste until he has squandered his whole patrimony and ends his melancholy career in the poor-house or the dungeon. We have, however, seen that even this will not save the wretched client, for he has already spent far more than he ever could have received originally by inheritance, and far more than he could have gained by gifts pitched in in bulk—like the poor colored brother's potatoes—through the window.

We have therefore gone over the case anew, and have learned that enormous electrical currents are constantly passing between the earth and the sun, with practically no resistance, and this irrespective of any hypothesis, actual or possible; and these facts have solved at the outset one of the greatest conceivable difficulties,—to wit, that of the transmission through space of such essential currents. Turning our attention to the more recent advances in electricity and the arts of electrical construction, we have found that induction machines, as contradistinguished from the older friction machines, operate in a manner strongly suggestive of the rotation of a planet through space, and we learn that the electrical potential of the air overhead increases constantly by an enormous multiplying number as we ascend, proving great electrical action in the regions immediately surrounding the earth, and which we have called the terrestrial electrosphere. We have also found that sun-spots and solar storms and other disturbances are at once reflected in our earth-currents, and are followed immediately by great electrical disturbances here and by extensive auroral displays at night. Experiment shows that similar auroral displays may be produced with an electrical machine by interruption of the current leading to its principal condenser, thus demonstrating that the currents are *from* the earth to the sun, and not the converse. We have also found that while the solar atmosphere is largely composed of hydrogen gas, that of the earth and other planets is largely composed of oxygen, and that these gases, the constituents of water, are separately disengaged at the opposite electrical poles by the electrolytic action of a powerful current of electricity applied to the decomposition of aqueous vapors, in accordance with the established electrical law that any fluid which will transmit a current may be decomposed by it; hence we learn that our interplanetary space contains attenuated aqueous vapors, which we have also learned to be true from other sources. As our other planets, as well as the earth, are found to be surrounded with an atmosphere of dilute oxygen, and with aqueous vapors suspended in it, we know that their action upon the sun must be similar to that of the earth, and that the congeries of planets thus unite in their supply of electricity to the sun in constant and enormous currents. Examining now the effects of passing powerful electrical currents through a compressed envelope of hydrogen gas surrounding a conductor, we find that great heat ensues, that the hydrogen becomes highly incandescent, and that the metallic nucleus within is raised to an extremely high temperature, and we also observe the same effects when the current is transmitted through the separated carbons of an electrical arc light. We have thus accounted for the constant supply of the energy which, transformed into light and heat, as in the last-mentioned experiments, the sun pours forth perpetually into space. We have also learned that electrical induction machines derive their electrical currents from the surrounding air, and also that no electricity can be generated in, or transmitted through, a vacuum, and hence we learn that the planets, by the rotation of their electrospheres in contact with

CONCLUSION. 345

the attenuated vapors of space, generate these powerful electrical currents with which the sun is supplied, and that the sun merely restores to the ocean from which, in another form, it was abstracted the light and heat which he emits, and that, instead of all being wasted except that which falls upon the planets, in fact that is the only part which actually, in one sense at least, is wasted: all the rest is deposited in bank, but that is "spent." The important generalization is thus arrived at, that the true source of solar energy is to be found in the attenuated vapors of space, and that the mode is that of the generation of electricity by the rotating planetary electrospheres, its transference through the aqueous vapors of interplanetary space to the sun, its passage under resistance through the compressed hydrogen envelope, its transformation there into light and heat, and its final emission or backpouring into space again. The molecular motions which give rise to light and heat in their passage through the vast distances of space are finally retarded by and disappear as radiated energy in the restoration or increase of the intermolecular tension of the vapors of space, and these processes continue, and must continue, to all eternity, if the sun exists and his planets continue to revolve in orderly circuit around him. If there be any permanent degradation of energy, it must be with reference to the total volume of infinite, or at least indefinite, space, and not with reference to the relatively minute spark of fire which we call the sun. We have also learned that the moon's electrosphere is repelled by that of its neighbor, the earth, and that whatever vapor and atmosphere it may have can exist only on its opposite side; and we have also learned that, by reason of the moon's peculiar axial rotation with reference to the earth, any other arrangement of the lunar moisture and air, even if such were possible, would have absolutely prohibited all life on that subordinate planet at any stage of its existence whatever. We have applied the above principles to the fixed stars, and have learned that, by the same law, the resplendent star itself is proof conclusive that it, too, must have planets rotating around it, and that these planets must have an oxygen atmosphere and clouds of aqueous vapor like our own. We have interpreted the double and multiple stars, and, by an extension of the same law, explained their frequently contrasted or complementary colors. The new stars which blaze up in sudden conflagration and then die out have no secrets when this new light is turned upon them; they, too, are but the faithful followers of the law; and the temporary and variable stars likewise fall into their appropriate categories and obediently move on with the procession. The comets,—the banner-bearers of the sidereal hosts,—which from the earliest ages have defied science to read their cabalistic legend, find it now "writ large" and in plain English. Even the meteorites, the cosmical dust, the unorganized *débris* of space, are found to be amenable to the same law. When we turn in wider gaze to spy out the fantastic nebulæ on the very outer fringe of visible things, after we have separated out the star-clusters and organized galaxies of suns, we apply our touchstone to the irresolvable gaseous nebulæ, and lo! their mystery dissolves at a touch. We have even been able to picture the processes of the creation of solar systems and whole galaxies of suns in which the same law finds scope, and by its infinite and harmonious extension we learn that nature moves with a comprehensive plan, and is uniform in her infinite variety and eternal in her ceaseless activity. We have been told that—

"The poem of the universe
No rhythm has nor rhyme;
Some god recites the wondrous song,
A stanza at a time."

But it is all a mistake; the loftiest strains which ever inspired the soul of Mozart or of Beethoven had not the ineffable harmony, nor the sweetest songs of the greatest poets the perfect rhyme, ever repeated and ever varied, of the universe. Its orderly progress is like the onward movement of a mighty army, and there is but one grand commander, "but one God," and Nature, that showeth forth his handiwork, "is his prophet." We have found that the "course of nature," the eternally youthful mother, is the same, whether in spinning a tendril in the garden, in weaving a whirlwind in the atmosphere, or in elaborating from the universal vapors of primordial space a solar system or a galaxy. And it is not a convulsive, spasmodic nature that we find; we do not love to associate great explosions, cataclysms, the destruction of worlds, or the extinction of suns with our ideas of nature. These seem not to be of nature. The nature we love is the gentle mother, uniform in her operations, kindly in her ways, beneficent in her results; the nature of the rain, the sunshine, seed-time and harvest and the sprouting seed again; ever patient, ever responsive, but in all as firm and steadfast as the foundations of eternity itself. So we have found her. We have assumed nothing; we have observed and endeavored to deduce from observation her systematic plan, for this is the voice of her law, "the same yesterday, to-day, and forever." To quote the words of Matthew Arnold, from out the darkness of the past we seem to hear her say,—

"Will ye claim for your great ones the gift
To have rendered the gleam of my skies?
* * * * * *
Race after race, man after man,
Have thought that my secret was theirs,
* * * * * *
—They are dust, they are changed, they are gone!
I remain."

REFERENCE INDEX OF AUTHORITIES CITED.

Appleton's Cyclopædia, pp. 21, 48, 49, 52, 56, 107, 131, 134, 148, 155, 156, 159, 162, 168, 181, 188, 200, 207, 262, 264, 267, 270, 294.
ARGYLE, 319.
ARNOLD (Matthew), 348.
AUGUSTINE (Saint), 329.
AYRTON, 77.
BALL, 9, 28, 34, 35, 39, 41, 48, 51, 54, 58, 61, 63, 79, 82, 128, 139, 158, 163, 170, 193, 206, 207, 216, 239, 241, 243, 245, 256, 266, 270, 272.
BEETHOVEN, 347.
Bible, 308, 327, 329, 330, 332, 333, 334, 337–340.
BODE, 287.
BRAHE (Tycho), 179, 342.
British Association, 206.
BUFFON, 21.
BYRON, 152.
CARRINGTON, 59, 75.
CLARK, 258.
COPERNICUS, 80.
CROOKES, 232, 297, 298.
CROWELL, 28.
D'ARREST, 257.
DARWIN (Charles), 28.
DEWAR, 213.
DRAPER (Dr.), 4, 7, 214, 217.
DULONG, 215.
DUNKIN (Prof.), 133, 159, 163.
Egyptian cosmogony, 316.
"Electrical Review," 85.
"Electricity in the Service of Man," 70, 74, 77, 83, 90, 91–94, 95, 105, 132, 176, 225, 233.
EMERSON, 248.
English version of Bible, 311.
Ethiopic sources, 316.
FARADAY, 123, 132, 227.
FERGUSON, 132.
FLAMMARION, 9, 238, 263, 305.
FLEMING (Prof. J. A.), 83.
FLIGHT (Dr.), 232.
FONTANELLE, 24.
FOSTER (Prof.), 78.
FOWNES, 215, 216.
FRAUNHOFER, 9, 87, 153.

GATHMANN (Prof.), 135.
GEIKE, 28.
GOETHE, 145.
GROOMBRIDGE, 63, 246, 249.
GUILLEMIN, 273, 285, 304, 305.
GUYOT (Prof. Arnold), 316, 319, 325.
HALE (George E.), 58.
HAMILTON, 124, 227.
HAUSEN, 122.
HELMHOLTZ, 9, 21, 23, 28, 31, 140, 288.
HERSCHEL (Alexander), 140.
HERSCHEL (Sir John), 229, 304.
HERSCHEL (Sir William), 9, 35, 58, 80, 148, 199, 239, 257, 258.
HERTZ, 79.
HIND, 257, 258.
HODGSON, 75.
HOLTZ, 94.
HUGGINS, 9, 55, 61, 79, 109, 158, 163, 181, 205, 212, 213, 214, 216, 217, 235, 254, 255, 258.
HUYGENS, 265.
Indian sources, 316.
JANSSEN, 49.
JOUVENCEL (De), 318.
KANT, 35.
KELVIN (Lord), 38.
KEPLER, 80.
KIRCHER, 329.
KIRCHHOFF, 53, 77.
LANGLEY, 9, 29, 33, 48, 58, 113.
LAPLACE, 35, 269, 275, 278, 279, 280, 309.
LIVEING (Prof.), 213.
LOCKYER, 49, 285, 305.
LOOMIS, 108.
LYELL (Sir Charles), 28.
"Mankind: their Origin and Destiny," 314.
McCOSH (Dr.), 317, 318.
McGEE, 28.
MANN, 334.
Masoretic pointing, 311.
MAYER, 21.
MELCONI, 149.
MILLER, 9, 26, 122, 158, 200, 204, 248.
Mosaic narrative, 310, 337-340.
MOSES, 313, 315, 330, 332, 334.
MOTT (A.), 39.
MOZART, 347.
MYER (Gen. A.), 9, 55, 56.
NEWCOMB, 9, 34, 270.
NEWTON, 21, 80, 228, 241, 280.
NICHOL, 9, 80, 164, 188, 238, 262, 263, 265, 278, 279.
PERRY, 77.
PETIT, 215.
PICKERING (Prof.), 258.
"Popular Science Monthly," 57, 113.

PROCTOR, 9, 24, 27, 35, 36, 37, 45, 46, 47, 51, 75, 78, 80, 97, 99, 108, 111, 145, 156, 157, 159, 166, 179, 182, 184, 191, 199, 200, 201, 204, 206, 207, 212, 220, 224, 231, 232, 234, 237, 253, 255, 258, 299, 302.
PTOLEMY, 342.
PYTHAGORAS, 329.
RAWLINSON (Prof. George), 359.
ROSSE (Lord), 9, 188, 255, 261, 266, 279.
ROUGÉ (M. de), 316.
ROWLAND (Prof.), 61.
RUPERT (Prince), 295.
SALISBURY (Lord), 38, 69.
SCHIAPARELLI, 200.
SCHMIDT (Dr.), 258.
SCHRÖTER, 134.
SCHUSTER (Dr.), 79.
SECCHI, 156, 157.
SEEBECK, 149.
Septuagint, 334.
SIEMENS, 21, 36, 37, 53.
SMYTH (Admiral), 163.
SPENCER (Herbert), 270.
STEWART (Balfour), 7, 140, 141, 142, 145, 146, 152.
STRUVE (O.), 257.
TAIT, 38, 204.
TENNYSON, 198, 268.
THEOPHILUS, 329.
THOMSON (Sir William), 25, 26.
TOEPLER, 95.
TYNDALL, 9, 104, 123, 146, 148, 149, 227.
UPHAM, 28.
URBANITSKY, 9, 70.
VOGEL, 170.
VOSS, 94, 233, 290.
WATERSTON, 21.
WELDON (Charles), 347.
WILSON, 123, 227.
WIMSHURST, 94, 132.
WOLCOTT (Prof. C. D.), 28.
WOLF, 107.
WRIGHT (Arthur W.), 52.
YORK (J. F.), 316.
YOUNG (Prof. Charles A.), 9, 53.

CLASSIFIED INDEX OF SUBJECT-MATTER.

ASTRONOMY.
Largely an empirical science, hitherto, 9.
New light on the phenomena of, 68, 250, 341.
Review of subject-matter of the present work, 341-348.
Speculative, excluded, 341.
Interpretation of the mysteries of, 348.

ATMOSPHERE.
Atmosphere of sun composed principally of free hydrogen, 39, 61.
Free oxygen the characteristic element in earth's atmosphere, 39.
Mott's theory to account for absence of hydrogen in earth's atmosphere untenable, 39-44.
No theory, hitherto, has accounted for the solar hydrogen, 44.
Aqueous vapors in planetary atmospheres, whence derived, 46, 62.
Aqueous vapors diffused through interplanetary space, 46, 65.
Aqueous vapors diffused through interstellar space, 65.
Composition of the terrestrial atmosphere, 47.
Composition of the solar atmosphere, 48.
Composition of the planetary atmospheres, 62.
Aqueous vapors around the sun, 62.
Two grand categories of heavenly bodies, one with atmospheres characterized by free hydrogen and the other with atmospheres characterized by free oxygen, 62.
Atmospheres, either electrically positive or negative, of hydrogen or oxygen, universal for all the bodies of space, 65.
Solar and cometic bodies have atmospheres of the hydrogen class, highly heated; planetary atmospheres are of the oxygen class, and are cool, 66.
Solar and planetary atmospheres are mutually correlated, and produced by disassociation of the elements of aqueous vapors, 67.
"No sun no planets; no planets no sun," 69.
Rapid increase of electrical potential as we ascend through the earth's atmosphere, 74.
Its significance, 74, 75.
Sun-spots, terrestrial electricity and magnetism, and auroras, connected with one another, 77.
A material medium, besides the luminiferous ether, exists between earth and sun, 81.
The medium consists of attenuated aqueous vapors commingled with other vaporized elements, 81.
The processes of formation of solar and planetary atmospheres from these vapors, 82, 308.
Incandescence of solar and cool state of planetary atmospheres explained, 83-85.

354 CLASSIFIED INDEX OF SUBJECT-MATTER.

ATMOSPHERE—(Continued.)

Contraction and expansion of sun's semi-vaporous condensed nucleus a self-compensating mechanism for the regulation of his light and heat, 88, 106.
Identity of atmospheric aurora and electrical brush-light discharge, 90, 91.
Rotating electrosphere of the earth, 96.
Dimensions of, 96.
Resistance of atmosphere considered, 97, 100.
Principles concerned in the generation and maintenance of atmospheres, 100-106.
Currents in space; their influence on planetary and solar electrospheres, 106-107.
No visible atmosphere on the moon, 122.
Atmosphere and aqueous vapors must exist on the moon's surface, but can exist only on opposite side, 123.
Lunar atmosphere and axial rotation considered with reference to "Argument of Design," 122-128.
Habitability of the other planets, 128-136.
Atmosphere of Mars analyzed and computed, 130-132.
Atmospheres of Jupiter, Neptune, the moon, etc., 132.
Method of computing the atmosphere of any known planet, 131-134.
Estimation of oxygen in different planetary atmospheres, 133.
A slight libration of the moon's atmosphere around its margin produced by counteractive angular effect of solar attraction and repulsion of the earth's electrosphere, and its result, 133-136.
Vegetation said to have been observed on lunar surface at margin of this libration, 134-135.
Aqueous vapors of space considered with reference to thermal light of the sun, 147.
Spectroscopic analysis of atmospheres of the stars, 156-161.
Interpretation of complementary colors of double stars, 163.
Mutual repulsion of similarly electrified atmospheres, 124, 166-167.
Variability of regularly variable stars produced by dynamic action of their planets, 168.
Atmospheres of temporary stars, "suns in flames," 195.
Effect upon planetary atmospheres of our system should our sun become such a "new star," 196-198.
Atmospheres of comets, 205, 212.
Atmospheric repulsion of sun and comet, 210.
Atmospheric attraction between planets and comets, 211.
Cyanogen as an element of cometic atmospheres, 216, 218.
Decomposition of cyanogen into non-toxic substances by contact of a comet with a planetary atmosphere, 218-219.
Temperature of cometic atmosphere, 218.
Repulsion of cometic atmosphere by the sun's electrosphere, 231, 235.
Development of planetary atmospheres during coalescence of ruptured convolutions of a spiral nebula into spheres, 291.
The attenuated vapors of space, 297-298.
The square-shouldered aspect of Saturn's atmosphere, first noticed by Herschel, explained, 302. (See also Fig. 4, page 124.)
Barometric pressure of earth's atmosphere highest in the temperate zones; its interpretation, 303.
Application of same principle to sun-spots, 303.

CLASSIFIED INDEX OF SUBJECT-MATTER. 355

ATMOSPHERE—(Continued.)

Should present atmospheres be conceived to be obliterated, new planetary and solar atmospheres would be generated precisely similar to those which now exist, 308–309.
Solar light and heat would again be re-established, 309.
Atmospheres in their characteristic elements all due to electrolytic decomposition, 343, 344.

BIOLOGY.

Compared with astronomy, 10.
Splendid advances in, during past few years, 15.
Laws of, as related to those of astronomy, 247.
Mosaic cosmogony as related to, 320.
Order of succession in the introduction of life, according to the Mosaic narrative. (See latter title in Index.)

CHEMISTRY.

Hydrogen of solar photosphere and chromosphere, 39.
Oxygen in earth's atmosphere, 45–47.
Chemical elements in the sun, 47, 61.
Absence of free oxygen in the sun, 47, 69.
Absence of free oxygen in comets, 62.
Elements found in comets, 62, 212, 218.
Olefiant gas in comets, 207, 232.
Hydrogen, carbon, sodium, and cyanogen, 213, 214.
Carbon and hydrogen compared, 214, 217, 260.
Reactions of cyanogen, 217.
Decomposition of cyanogen by contact of comets with a planetary atmosphere, 218, 219.
Gases occluded in meteorites, 232.
Laws of crystallization, 247.
Chemistry of gaseous nebulæ, 254–262.
Nitrogen, hydrogen, and (most probably) oxygen in all gaseous nebulæ, 254.
Possibly a more elemental condition of gases in nebulæ, 259.
Ammonium a *hypothetical* inorganic radical, 259.
Bright-line spectrum of gaseous nebulæ, 267.
Chemical changes during progression of spiral nebulæ, 287–292.
Oxidation of terrestrial mass during coalescence, 292.
Phenomena of nature, 299, 341.

COMET.

Some of the phenomena of, can only be accounted for by electricity, 7.
Hydrogen and nitrogen in comets, but no oxygen, 62.
Description of the phenomena of comets, 200, 203, 210.
Trains of meteors follow track of comets, 203–204, 206–207, 232.
Formation of envelopes and tails, 205, 220.
Olefiant gas in comet and meteorite, 207, 232.
Electrical repulsion of comets' tails, 208, 225–231.
Mass and tenuity of comets, 209, 223.
Opposite electrical polarity of comets and planets, and similar polarity of sun and comets, 211, 233, 236.

356 CLASSIFIED INDEX OF SUBJECT-MATTER.

COMET—(Continued.)
Spectra of comets, 213.
Hydrogen compounds in comets, 213.
Temperature of cometic nucleus, 218.
Reversal of polarity of comet by contact with a planetary electrosphere, 233–234.
Comets most frequently without tails, 222, 281.
Interpretation of the phenomena of comets, 235.
Repulsion of comets' tails illustrating phenomena of gaseous nebulæ, 280.
Many comets transcend that of Newton in dimensions of their tails, 281.
Origin of comets by excessive repulsion from the nebular matter of a forming solar system, 289.
Phenomena of comets in accordance with universal laws governing celestial bodies, 346.

COSMOLOGY.
According to previously accepted views the visible order of creation must result in a final failure, 18.
Possible termination of present cycle of terrestrial life and possible renewal, 198.
Solar systems not necessarily individual creations, 165.
The word "creation" as rendered in our version of the Bible, 320.
Mosaic narrative (see this title in Index), 337–340.
Mosaic cosmogony does not exclude prior material space, 320.
Original creation out of nothing forms no part of the Mosaic or of other primitive cosmologies, 320, 329, 330.
Nebular hypothesis not in accordance with Mosaic account of creation, 327.
Knowledge of cosmology among the ancients, 328, 329.
Ancient Egyptian cosmogony, 316.
Ancient Syriac cosmology, 330.
Second Mosaic narrative (the garden of Eden), 334–336.
Literal translation of the Mosaic record of the creation, 337–340.
Review of the system of cosmology embraced in the present work, 341–348.
The harmony of nature's operations, 341.
Universal cataclysms contrary to nature, 347, 348.

ELECTRICITY.
Electrical connection between earth and sun, 7.
Mere currents can play no part in the grander operations of nature, 8.
Repulsion by the sun of the solar corona, 55, 61.
Electricity, the universal source of repulsion, compared with gravity and affinity, the universal sources of attraction, 70.
Electricity considered with reference to solar energy, 70, 343.
Electrolysis, 70.
Laws of electricity, 70.
Currents constantly passing between earth and sun, 75.
The same considered in detail, 75–76, 80, 343.
Velocity of these currents equal to that of light, 77.
Cannot pass through vacua, 81.

CLASSIFIED INDEX OF SUBJECT-MATTER. 357

ELECTRICITY—(Continued.)

Heating effect of electrolyzing current, 83, 344.
Arc lamp, 83–84.
Intense heat produced by current under water, operating through a hydrogen envelope surrounding a conductor, 85.
Electrical induction machines described, 88–95, 344.
Their resemblance to rotating planetary electrospheres, 96, 345.
Mutual repulsion of similar electrospheres, 123–125.
Analogy of reflex nervous system with electrical circuit, 136.
Phenomena of variable stars due to more or less concentrated electric currents from their encircling planets, 175.
Variation in constitution of, and currents in space affect the planetary generation of electricity, 188–192.
Electricity between adjacent solar systems, 194.
Electrical repulsion of the tails of comets, 211, 235.
Electricity as an element in development of nebulæ, 284–286.
Electrical repulsion operates to drive off the matter of future comets from condensing nebulæ, 289.

HYPOTHESIS. (See Theory.)

No adequate hypothesis, hitherto, to account for continuance of solar energy in the past, 17.
General statement of Laplace's nebular hypothesis, 12.
The nebular hypothesis has not been proved, 35, 270–278.
What it requires for its basis, 97, 274–276.
Correct basis for hypothesis of solar energy, 141–144, 286.
Nebular hypothesis considered in detail, 268–278.
Contrast of nebular hypothesis with the present work, 306.
The Mosaic cosmogony, 308.
Nebular hypothesis deals only with aggregations, 309–310.
The cosmogony of Genesis more scientific, 310.
Origin of Mosaic narrative, 310, 329–330.
Egyptian cosmogony, 316.
Different hypotheses reviewed, 342.
All prior theories insufficient to account for the facts, 342.

LAW, NATURAL.

Some general law must control astronomical phenomena, 7.
But few fixed, controlling laws in nature, 14.
Natural laws eternal in their operation, 18.
Supremacy of natural laws, 100.
Gravitation cannot control star-drift in space, 64.
Universality and harmony, but not identity in the results of the operation of these laws, 68.
"A more wonderful law of harmony than those of Copernicus, Kepler, and Newton," 80.
Indefinite approaches often prelude great discoveries, 80.
Laws of repulsion and attraction, 124–127.
Harmony among all the solar systems, 145, 153.
Sphere of effective control under gravity, 241.
Universality of gravitation has been doubted, 241–242.
Demonstration that gravity cannot control universally, 243–245.
Proportionate and aggregate attractions between systems, 244.
Stars traverse space without reference to law of gravity, 246.

358 CLASSIFIED INDEX OF SUBJECT-MATTER.

LAW, NATURAL—(Continued.)
A higher law of movement indicated, 247, 249.
Comparison with the natural laws of biology, 247.
Laws operate constantly, but only manifest change at intervals, 248, 283.
The drift of stars through space, 249.
Interdependence between all created systems, 250–252.
Astrology: its abandoned beliefs considered, 261.
Attraction and repulsion naturally correlated, 280.
Bode's empirical law interpreted by development of the solar system from a spiral nebula, 287.
Arrest of moon's axial rotation, 293.
Laws of Laplace, etc., 294.
Laws of movement in the development of solar systems, 298.
Basis of human knowledge, 299.
Interpretation of the laws of nature, 306–307.
Operation of same laws which produced our solar and planetary atmospheres would reproduce similar ones if these were destroyed, 308.
Universality of natural laws, 347, 348.

MOSAIC NARRATIVE.
Moses fully acquainted, by initiation into the priesthood, with the sacred knowledge of the Egyptians (the Hebrews were not), 310.
The Mosaic record more scientific than the Nebular hypothesis, 310.
Improperly rendered from the original in our version, 310.
Full and correct translation not then possible, 310.
Hebrew a root-language, and not original or inspired, 311.
Indefiniteness of translation in our version illustrated, 311–312.
Importance of accurate rendering of the words of the original, 313–314.
Cannot be interpreted by writings made long subsequently, 313.
Correct basis of a true rendering, 314.
Use of the important words of the original, 315.
Jehovah not directly mentioned in the narrative; the work was performed by specially energized natural forces operating under guidance of a higher power, 315–316.
Ancient Egyptians believed in one supreme God, 315.
Also the Aryans of prehistoric times, 316.
The cosmogony of the Egyptians, 316.
Dr. McCosh on the delegated forces of God, 317–318.
The word which is translated "rested," 318, 340.
Analogy of volcanic action with work of creation, 318.
Professor Guyot on the meaning of "God rested;" the forces of nature came to a state of equilibrium, 319.
Duke of Argyle on the processes of creation around us daily, 319.
The words "created" and "made," in verse 3, chapter ii, not properly rendered; popular misconception based on this imperfect rendering, 319.
Signification of the words BRA, OSH, and IEI, 320–323.
Separation of waters to two opposite foci, with attenuated space between, 324, 325, 329.
The above separation hitherto misunderstood, 325.
Better known to the ancients, 328, 329.

CLASSIFIED INDEX OF SUBJECT-MATTER. 359

MOSAIC NARRATIVE—(Continued.)
Song of the Three Holy Children, the Psalms, Theophilus, and St. Augustine, on the separation, 329.
Introduction of vegetable life prior to appearance of free oxygen in earth's atmosphere, 323-326.
Jeove as contradistinguished from Aleim, 327.
Mosaic cosmogony based on prior attenuated matter of space, 327.,
Astronomical knowledge of ancient peoples, 329.
Table of root-meanings of words used in the narrative, 330-333.
Some portions of the second narrative examined, 333-336.
   NOTE.—The second narrative bears the unmistakable impress of its sacred Egyptian derivation; the temptation is pictorially represented on the walls of the temple of Medinet-Abou, at Thebes, which dates from the eighteenth dynasty, while Moses was contemporary with the nineteenth. Joseph entered Egypt during the Hyksos period preceding the eighteenth. (Rawlinson, "Ancient Egypt." See also his "Ancient Religions," for Egyptian monotheism, last three pages of chapter i.)
Popular need of a more accurate translation of the earlier Scriptures, 336.
The narrative of creation literally translated, 337-340.
Order of the successive introductions of life, according to the Mosaic record: 1, land plants; 2, marine vegetation (necessary for sustenance of 3); 3, lower forms of marine life; 4, reptiles; 5, *birds* (between reptiles and the mammalia); 6, mammals; 7, mankind, male and then female, 338, 339.

NEBULA (Gaseous).
Hydrogen and nitrogen in, 62, 216.
Elongated nebula in Sobieski's Crown, 189.
Gaseous nebulæ affected by currents in space, 189.
Oxygen in gaseous nebulæ, 216.
Distribution of nebulæ in space, 237-238, 262, 264.
Herschel's arrangement of, in progressive series, 239.
Great composite nebula in Orion, 240, 255.
Gaseous nebulæ described, 253.
Spectroscopic analysis of, 254-258.
Changes in form of gaseous nebulæ, 256-258.
Reversion of a small planetary nebula, 258.
Progressive changes in nebulæ, 258-259, 267.
Analysis of drawings of gaseous nebulæ of Lord Rosse, 261-262, 265.
Typical forms of non-systemic nebulæ, 263.
Crab nebula, 265, 285.
Number of gaseous nebulæ already recognized, 265.
Spiral figure a characteristic, 265, 266.
All spectra of gaseous nebulæ show bright lines, 267.
Development into solar systems, 267, 283.
Spiral nebula in Canes Venatici, 273.
Series of spiral nebulæ illustrating progressive advances, 279.
Types of development, frontispiece and legend beneath.
Comparison of spiral nebula with a jet of water, 285.
Comparison with tail of a comet under rotation, 285.
Development in accordance with general astronomical laws, 346.

NEBULA—(Continued.)

Convolutions of spiral nebula pyriform, 293.
Origin of nebulæ from the matter of space, 295.
Production of planetary nebulæ by mutual repulsion, 301-302.
Distances of gaseous nebulæ hitherto overestimated, 303, 304.
Each spiral nebula develops into a single solar system, 304.
Spiral character of many apparently globular nebulæ revealed by telescopes of adequate power, 304-305.

PLANET.

Those of our own system resemble each other, 45, 67.
Jupiter's body covered with clouds and invisible to us, 45.
Saturn, Venus, Mars, 45.
Surface of Mars clearly marked, rarely concealed by vapors, 45-46.
The planets of our own solar system the only ones visible to us, 63.
Every self-luminous star must have planets rotating around it, 63.
Some solar systems may have a single planet, 67, 171, 302.
How planets generate electricity from space, 88-89.
No visible atmosphere or aqueous vapor on moon, 122-136.
Center of gravity of moon apparently displaced, 122.
The atmosphere of Mars, its constitution, 130-132.
Planets belonging to solar systems with double suns, 167-168.
Angular positions of planets regulate solar energy, 176.
Due to inclination of solar axis, 119-122.
Formation of planets from the convolutions of spiral nebulæ, 286, 289, 292.
Abnormalities of planets in our system accounted for, 286-287, 294.
Formation of planetary satellites and Saturn's rings, 292-293.
Formation of belt of asteroids, 294.

SOLAR ENERGY.

Our first investigations directed to phenomena of our own solar system, 8.
Successively extended to other bodies of space, 8.
Simple uniformly acting laws which control, 9.
Different theories of, hitherto in vogue, 17, 34.
Gradual degradation of, according to former theories, 18.
Primary error due to attributing solar energy to an original supply in the sun, 19.
In truth, it is derived from the rotation of the surrounding planets, 65.
Produced by electrical currents from planetary electrospheres, 83-86.
Experiment with hydrogen envelope in a pail of water, 85, 344.
Its production and permanent maintenance, 86, 88, 195.
Its mode of distribution, 139, 345.
The apparent waste not real, 140, 345.
Correct statement of the mode of production and distribution of all solar energy, 141-145, 344-346.
Discussion of the light and heat of, 147-152.
Due to planetary energy; evidence from the variable stars, 175, 346.
Great heat-wave of 1892, 193.
Illustration of solar energy, analogous to water-wheel, 251.
True final source of solar energy, 252, 345.
Nebular hypothesis with relation to, 268-274.

CLASSIFIED INDEX OF SUBJECT-MATTER. 361

SOLAR ENERGY—(Continued.)
Difficulties of nebular hypothesis, 274–278.
Spiral nebulæ incompatible with nebular hypothesis of, 273–278.
Splitting up of gaseous nebulæ by internal repulsion, 289.

SOLAR SYSTEM.
Belief, hitherto, in its early termination in eternal darkness, 18.
Constitution of our, 62.
Drifting through space, 63.
Suns and planets mutually correlated, 69.
Electrical connection between sun and planets, 79.
Only $\frac{1}{2320000000}$ part of sun's energy received by our planets, 139.
Solar system of variable star Mira, 177.
Operation of solar systems perpetual, 198.
No operative solar system could be built up from meteorites, 199.
Views expressed in this work contrasted with former theories, 250–251.
Development of a solar system from a spiral nebula, 279.
Genesis of solar systems from the primordial elements of space, 282.
Phenomena of the development of solar systems, 283.
Mode of development of a centripetal planetary solar system from a centrifugal spiral nebula, 286.
Mode of formation of the asteroids, 288.
Of comets, 289.
Disruptive force of repulsion in a gaseous nebula, 289.
Rupture of convolutions preparatory to formation of planets, 290.
Reversal of electrical polarity of ruptured convolutions, 290.
Coalescence into separate planets, 290–292.
Periodicity in the development of solar systems, 300.
Origin of single planet solar systems, 171, 302.

SPACE.
Estimated temperature of, 82.
Currents in, 106, 187–189.
Distribution of stars in space, 187.
Universal connection between all bodies of space, 250.
So-called "empty space," 295.
Tensions in space, 295.
Illustration from Prince Rupert's drops, 295–296.
Constitution of space, 297.
Unstable equilibrium, 297–298.
Apparently blank areas of space, 299.
Our present space eternal, 299.
The attenuated vapors of space the source of all created things, 299–300.
The domain and workshop of the Infinite, 307.
The last refuge of the human intellect, 307.

SPECTROSCOPE.
Absorption bands and bright-line spectrum, 155.
Spectroscope as used in investigation of nebulæ, 253.
Applied to great nebula in Orion, 256.
Bright-line spectra in all gaseous nebulæ, 267. (See Chemistry, Star, Sun.)

362   CLASSIFIED INDEX OF SUBJECT-MATTER.

STAR.
Distances of stars in space, 64, 244, 248.
Our sun a variable star, 75, 179.
Classification by their spectra, 156–158.
Betelgeuse, 159, 161.
Double stars, 162.
Double and multiple stars of complementary colors, 162–164, 176, 305.
Origin of double stars, 164, 167, 305.
Mizar, 165.
Interpretation of phenomena of double stars, 168.
Variable stars, 168.
Regularly variable stars, 169.
Algol, 169–173, 302.
Planetary system of Mira, 177.
Delta Cephei, 174.
Variability due to variable dynamic energy of planets, 119–122, 175.
Phenomena of temporary stars, 180–182.
Insufficiency of previous explanations of, 183–186.
True causes of, 187–196.
Temporary stars usually appear in certain parts of the heavens only, 192.
Star-clusters, 240.
Limits and structure of the Milky Way, 244.
How stars travel through space, 249.
Common brotherhood of all stars, 250.
Correct principles of interpretation and explanation of the phenomena of the stars, 346.

SUN.
Hitherto accepted belief that his energies are dying out, 18.
Chemical elements in the sun, 47.
Constitution and structure of the sun, 48, 61.
Prominences, faculæ, sun-spots, chromosphere, photosphere, corona, long streamers, solar nucleus, 48–56.
Sun-spots travel more rapidly across the solar face in proportion to their distance from his equator, 54, 59.
General Myer on sun's corona, 56.
Sun-spots described, 56–59.
Every sun must have planets to enable it to give out light and heat, 66.
Sun-spots and terrestrial electricity and magnetism, 75–76, 303.
Eleven-year period of sun-spots, 75.
Operative artificial sun; electrical experiment, 86–87.
Sun's gaseous or partially gaseous body a self-compensating mechanism to distribute and equalize his energies, 88, 106, 199.
Sun-spots considered with reference to angular positions of the planets, 107, 119–122, 155–156.
Origin and development of sun-spots, 107–122.
Our sun a variable star, 179.
Repulsion of sun's long streamers, 166, 280.
Cycles of life on the planets might be produced by successive increases and diminutions of sun's radiant energy, 197.
Repulsion of the tails of comets by solar electrosphere, 211.
Idea of a universal central sun untenable, 241.
Importance to mankind of a correct knowledge of the sun, 251.

CLASSIFIED INDEX OF SUBJECT-MATTER. 363

THEORY. (See Hypothesis.)
Various previous theories to account for solar heat and light, 19.
1, sun now giving out the heat imparted at its creation, 21.
2, that its volume is being consumed by combustion; 3, that its light and heat consist of currents of electricity; 4, that comets are the aliment of the sun; 5, that the supply is due to accretion by meteoric streams; 6, that it is due to molecular condensation from contraction of the sun's gaseous body; 7, Dr. Siemens's theory of disassociation of gases in space by sunlight and heat, centripetal suction at the solar poles, and recombination and centrifugal emission around the sun's equator, 21-22.
The above theories separately considered, 23-38.
Not sufficient, one or all, 39.
All fail, also, to account for the solar hydrogen, 39.

UNIVERSE.
Harmony throughout the universe, 68, 153, 341.
Classification of bodies which occupy the, 153.
Star-drift through space, 165.

www.ingramcontent.com/pod-product-compliance
Lightning Source LLC
Chambersburg PA
CBHW020236240426
43672CB00006B/547